Comparative Psychology of Invertebrate

Research in Developmental and Comparative Psychology
(Vol. 2)
Garland Reference Library of Social Science
(Vol. 1082)

Research in Developmental and Comparative Psychology

Gary Greenberg and Ethel Tobach, *Series Editors*

Behavioral Development
Concepts of Approach/Withdrawal and Integrative Levels
Edited by Kathryn E. Hood, Gary Greenberg, and Ethel Tobach

Comparative Psychology of Invertebrates
The Field and Laboratory Study of Insect Behavior
Edited by Gary Greenberg and Ethel Tobach

Comparative Psychology of Invertebrates

The Field and Laboratory Study of Insect Behavior

edited by
Gary Greenberg
Ethel Tobach

Routledge
Taylor & Francis Group

NEW YORK AND LONDON

First published 1997 by Garland Publishing, Inc.

This edition published 2012 by Routledge:

Routledge
Taylor & Francis Group
711 Third Avenue
New York, NY 10017

Routledge
Taylor & Francis Group
2 Park Square, Milton Park, Abingdon,
Oxfordshire OX14 4RN

First issued in paperback 2016

Routledge is an imprint of the Taylor and Francis Group, an informa business

Library of Congress Cataloging-in-Publication Data

Comparative psychology of invertebrates : the field and laboratory study of
insect behavior / edited by Gary Greenberg and Ethel Tobach.
     p.   cm. — (Garland reference library of social science ; vol.
1082. Research in developmental and comparative psychology ; v. 2)
Includes bibliographical references and indexes.
ISBN 0-8153-2196-1 (alk. paper)
     1. Insects—Behavior. 2. Insect societies.   I. Greenberg,
Gary. II. Tobach, Ethel, 1921– . III. Series: Garland reference library of social science ; v. 1082. IV. Series: Garland reference library
of social science. Research in developmental and comparative
psychology ; vol. 2.
QL496.C648   1997
595.7'051—dc20
                                                      96-12257
                                                        CIP

ISBN 13: 978-1-138-97128-8 (pbk)
ISBN 13: 978-0-8153-2196-5 (hbk)

On behalf of the Board of the T. C. Schneirla Research Fund, we dedicate this volume to the memory of Lester R. Aronson, a member of the group that developed the biennial meetings on which volumes in this series are based. He was an internationally respected scientist who, through his research and writing, contributed much to our knowledge and understanding of the development and evolution of behavior, which we are proud to have included in previous volumes in this series. He was an important guide to his students and colleagues, especially to the board in all its activities. Our loss of his keen, soft-spoken analyses of problems in science and everyday affairs will continue to be deeply felt.

T.C. Schneirla (1902–1968) in his office at the Department of Animal behavior, American Museum of Natural History, 1965. Photo by Charles Tobach.

Contents

Series Editors' Foreword

Research in Developmental and Comparative Psychology is a scholarly series of works dedicated to honor the contributions of T. C. Schneirla to comparative psychology, a discipline that crosses many boundaries in its concern with evolutionary theory. A consistent critic of anthropomorphism, Schneirla questioned vitalistic approaches that stressed only similarities in human and nonhuman behavior. He stressed the concept of integrative levels as an alternative approach that clarifies the continuities and discontinuities in the evolution of behavior. Most significant was the emphasis he placed on the study of developmental processes as an alternative to the contradictory (i.e., nature-nurture) hereditary and environmental explanations of the evolution of behavior. Schneirla's criticisms of ethology, based on the integrative levels concept, stand today as a valuable basis for criticism of sociobiology and evolutionary pschology. Schneirla's opposition to biological reductionism influenced research on the relationship between psychology and physiology. His approach/withdrawal theory, first formulated some 30 years ago, continues to form the basis of research and theory in the areas of behavioral development and emotional behavior.

Although he died in 1968, Schneirla is still recognized as one of the foremost theoreticians in comparative psychology, and the rebirth of interest in comparative psychology is evident in the work of many of his students and colleagues. The T. C. Schneirla Research Board convenes biennially to bring together the best writers on issues in comparative psychology. These sessions are cosponsored by Wichita State University, The American Museum of Natural History and the Graduate Center of the City University of New York.

The session on which this book is based was supported by a City of Wichita Mill Levy Grant, by Wichita State University and by the American Museum of Natural History.

Gary Greenberg
Ethel Tobach

Preface

When T. C. Schneirla became terminally ill, he was writing *Army Ants: A Study in Social Organization*. The book was completed by his last student, Howard R. Topoff, and was published in 1971. The papers in this volume on invertebrate behavior, predominantly ant behavior, are presented as a tribute to the man and to his theoretical and experimental contributions to our understanding of the development and evolution of behavior. (The editors define comparative psychology as the comparative study of the development and evolution of the behavior of all living organisms.) Themes developed by Schneirla in his research and theoretical formulations are evident throughout the volume.

Although he was not the first to use both laboratory and field techniques, his careful and insightful integration of laboratory and field techniques was a hallmark of his experimental approach. He broadened Wheeler's concept of trophallaxis by conceptualizing the reciprocal stimulation process that is illustrated throughout the experimental work reported herein as a central component of the social organization of the social insect.

At the time that Schneirla was most productive in his theoretical writings, the tradition of ethology was dominant in most European studies of comparative psychology. He rejected instinct, a central theoretical component of ethology, as an explanatory concept in behavioral research. Instead, he elaborated and gave new meaning to development as the focal process in understanding behavior. By doing that, he drew attention to the behavioral plasticity of developing organisms through all stages of their life cycles, and particularly highlighted the significance of this behavioral and developmental plasticity in the social insects. This aspect of the evolution and development of the social insects discussed in this volume speaks to the early contribution by Schneirla to this approach in the study of insects and of invertebrate behavior.

His emphasis on development also brought to the fore new questions, many of which are addressed in this volume. These questions center on the significance of different environmental processes in the integration of behavioral patterns evidenced in the ways in which the invertebrates and especially the social insects solve problems presented to them by the complexity of their abiotic and social environments.

Advances in technical instrumentation for research will be useful in reformulating these old questions in new and significantly constructive programs for responsible research. The theoretical contributions of Schneirla will continue to prove an important facilitation of those new research techniques.

Gary Greenberg
Ethel Tobach

Section I

Persistent Issues in the Comparative Study of Behavioral Evolution and Development

The Role of Controversy in Animal Behavior

Adrian Wenner

*What I saw when I looked at the famous duck-rabbit was either
the duck or the rabbit, but not the lines on the page—at least not
until after much conscious effort. The lines were not the facts of
which the duck and rabbit were alternative interpretations.*

(Thomas S. Kuhn, 1979, p. ix)

As a philosopher of science, T. C. Schneirla understood the important distinction between facts and their interpretation. He was clearly no stranger to controversy, as Piel pointed out in his introductory essays to a book honoring Schneirla (1970) and to the published version of the inaugural Schneirla conference (1984). His involvement in controversy—scientific, philosophical, cultural and political—is part of the reason that Schneirla "was not popular, not celebrated in the gatherings of psychology in his time. Schneirla was not in the mainstream" (Piel, 1984, p. 13). A pertinent example is the intellectual battle he waged with both the European ethologists and the American operant psychologists (Piel, 1970).

In retrospect one can well wonder why divergent views held by the ethology and the American psychology schools led to such intense controversy. That is because few realize that controversy, in part, is a recurring component in collective scientific research, albeit a terribly inefficient and quite unnecessary complication. We can appreciate this complication if we are willing, but another problem was recognized by Anderson (1988, p. 18): "To end controversies, scientists must first understand them, but scientists would rather do science than discuss it." More optimistically, Latour (1987, p. 62) stated: "We have to understand first how many elements can be brought to bear on a controversy; once this is understood, the other problems will be easier to solve."

To understand scientific controversy, we first have to understand how science operates, a topic normally given scant attention by scientists—just as Anderson emphasized. Schneirla clearly understood how science is actually more a process than a series of accomplishments. By contrast, biology textbooks extoll the accomplishments of scientists, largely ignore scientific process, and omit mention of controversies that may have preceded given accomplishments.

Psychologists, who receive extensive exposure to the history, methodology and philosophy of science, even as undergraduates, are usually surprised when they learn that these topics are nearly absent from the formal education of biology students in this country. Two decades ago I checked dozens of college catalogues across the country and found that courses in the above subjects were notably absent from undergraduate biology curricula, while nearly universally required for psychology undergraduates.

Physics is apparently in little better shape than biology, leading Theocaris and Psimopoulos to comment (1987, p. 597): "The hapless student is inevitably left to his or her own devices to pick up casually and randomly, from here and there, unorganized bits of the scientific method, as well as bits of *un*scientific methods."

The problem is not a new one. Ludwik Fleck (1935/1979) recognized the presumed disparity of approach between those in the "hard sciences" and those in the "soft sciences" when he wrote (p. 47): ". . . thinkers trained in sociology and classics . . . commit a characteristic error. They exhibit an excessive respect, bordering on pious reverence, for scientific facts." Neither did Fleck leave those in the "hard sciences" untouched; he wrote (p. 50) that the error of natural scientists consists of "an excessive respect for logic and in regarding logical conclusions with a kind of pious reverence."

Unfortunately, most ignore the great amount of accumulated thought (wisdom) that has been published in past decades, including notions about scientific process. While others have treated parts of that process in depth, a brief review of the overall collective process can be included here, summarized from more complete accounts published earlier (e.g., Wenner, 1989, 1993; Wenner & Wells, 1990).

Scientific Process

Unless science students are thoroughly inculcated with the discipline of correct scientific process, they are in serious danger of being damaged by the temptation to take the easy road to apparent success. . . . [They] should understand all the subtle ways in which they can delude themselves in the design of observations and the interpretation of data and statistics.(Branscomb, 1985, pp. 421, 422)

My interest in an analysis of this process began when Patrick Wells and I, along with our co-workers, became embroiled in a major scientific controversy in the mid to late 1960s (e.g., Wells & Wenner, 1973; Wenner & Wells, 1990). That controversy (the question of a "language" among honey bees) continues to this day.

We felt at the time that the bee language controversy should not have emerged from our test of that hypothesis. The relevant scientific community not only rejected our alternative interpretation but also ignored or summarily dismissed the anomalous results we had obtained, for the most part without even repeating the critical experiments that had yielded our results. Instead, Maier's Law (Maier, 1960, p. 208) prevailed: "If facts do not conform to the theory, they must be disposed of."

Consequently, we studied the history, philosophy, sociology, psychology and politics of science for two decades in an attempt to decipher what had transpired in the controversy that erupted as a result of our test of the language hypothesis (see Wenner & Wells, 1990). From that study, we recognized that scientific progress occurs (as indicated above) collectively and inefficiently by an unconscious group application of a definable method. We gradually formulated a diagram for our perception of this process (Figure 1).

The diagram is simple in principle. A new research trend begins (lower-right hand corner—*exploration* approach) when an individual recognizes (not merely observes) an important anomaly in nature while engaged in "normal science" (e.g., Kuhn, 1962/1970; Polanyi, 1958). Unconsciously, perhaps, that individual has moved from a

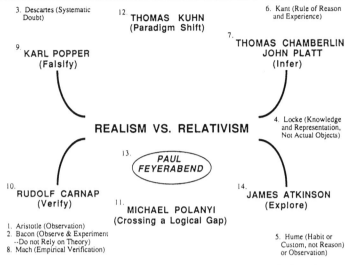

Figure 1. The collective scientific process *and how portions of it have been perceived through time, with each portion having had its advocates. For each sequence on any one collective research project, the complete process starts in the lower right-hand corner and progresses clockwise around the diagram, as the scientific community expands the scope of its inquiry (some steps may be omitted by practitioners). Movement around the diagram can stall (paradigm hold, see Figure 2), at which time progress plateaus. The numbers represent a chronology of contributions to formation of the diagram. See text for further explanation.*

state of "realism" ("knowing" what reality is) to "relativism" (an interpretation previously held to be "fact" is now suspect). The scientist then "creates an image" (Atkinson, 1985), forms an alternative explanation for evidence at hand and attempts to convert others to the same point of view.

If others can be convinced of the new interpretation, the scientist's view is reinforced (incipient "vanguard science"—Fleck's term, see below), moves back to the "realism" mode and attempts to verify the results (*verification* approach), a necessary but not sufficient part of the overall scientific process. When others can verify the results, many scientists can become committed to the new interpretation. A new research emphasis and protocol may then arise in a broader portion of the scientific community ("vademecum science"— Fleck's term, see below; also later termed "normal science" by Kuhn).

Unfortunately, much of animal behavior research during the past three decades has relied on verification alone (a partial view of the scientific process and only a portion of the "logical positivism" or "logical empiricism" school). That is, testing each hypothesis was not considered necessary in much of animal behavior research whenever a large body of evidence supported a given hypothesis (e.g., Wenner & Wells, 1990, pp. 204, 234). Animal behaviorists are not alone; scientists in general are reluctant to test their hypotheses (e.g., Mahoney, 1976). Such an attitude, of course, was responsible for "cold fusion" and other debacles in chemistry and physics (e.g., Asimov, 1989; Huizenga, 1992; Rousseau, 1992; Taubes, 1993).

During the 1960s and 1970s, researchers in ecology (e.g., critiques by Dayton, 1979; Loehle, 1987) and in psychology (e.g., critique by Mahoney, 1976) adopted another rather narrow approach (moving further clockwise around the diagram); they insisted that research be molded into an appropriate "null hypothesis" (*falsification*) protocol. The implicit rationale: If a premise cannot be proven false, then it is likely true or has some "probability" of being true ("realism" school).

In part, Thomas Kuhn's influence (anomalies emerge and eventually hypotheses become rejected, even without application of the formal null hypothesis approach) gradually forced psychologists to abandon their former comfortable stance. However, Schneirla had earlier perceived the weakness of that "working hypothesis" (e.g., Chamberlin, 1890/1965) approach, as phrased by Tobach (1970, p. 239): "He rejected logical positivism and operationism as bases for scientific inquiry and opened the way to a dynamic, holistic approach based on process." Ecologists continue to demand conformity to the null hypothesis approach (Dayton, 1979; Loehle, 1987).

In 1890, Thomas Chrowder Chamberlin (upper right corner of the diagram) recognized the weakness of an overreliance on verification ("ruling theory," in his terms) and/or on falsification (attempting to falsify a "working hypothesis," as phrased by Chamberlin). He advocated application instead of "The Method of Multiple Working Hypotheses" (*inference* approach), employing "crucial" experiments designed to provide mutually exclusive results. In that approach, scientists continually pit hypotheses against one another and attempt to falsify all of them during experimentation. After additional evidence is in, new alternative hypotheses are generated that might explain known facts and other pertinent information (e.g., Platt, 1964).

Only rarely does one find a scientist who can move from one approach to another with ease, as Claude Bernard, Louis Pasteur and Schneirla seem to have done, and as Feyerabend (1975) suggested in his famous phrase, "anything goes." Duclaux (1896/1920), biographer of Pasteur, recognized another root problem with respect to "objectivity" and experimental design for those who attempt to use standard procedure when he wrote: "However broadminded one may be, he is always to some extent the slave of his education and of his past." Four decades later, Fleck (1935/1979, p. 20) formed much the same conclusion: "Furthermore, whether we like it or not, we can never sever our links with the past, complete with all its errors."

Bernstein summarized succinctly the dichotomy between realism and relativism (1983, p. 8):

The relativist not only denies the positive claims of the [realist] but goes further. In its strongest form, relativism is the basic conviction that when we turn to the examination of those concepts that philosophers have taken to be the most fundamental . . . we are forced to recognize that in the final analysis all such concepts must be understood as relative to a specific conceptual scheme, theoretical framework, paradigm, form of life, society, or culture.

Ludwik Fleck, Overlooked Sage

Ludwik Fleck was a medical doctor in Poland in the 1930s and an expert on syphilis and typhus, expertise that kept him from being killed in concentration camps during WWII. While earlier studying the history of changes in attitude toward syphilis through time, he recognized the tentative nature of scientific "fact" and published a monograph in 1935, entitled *Genesis and Development of a Scientific Fact.*

Many of the points covered therein parallel notions advocated by those in the Schneirla school.

Thomas Kuhn wrote a foreword to the 1979 translation of Fleck's book, in part to acknowledge his indebtedness to the work (having read it in German before publication of his own classic 1962 work) and in part to explain that the volume contained much that he had missed earlier. Kuhn wrote (1979, p. x): "Though much has occurred since its publication, it remains a brilliant and largely unexploited resource." Kuhn also recognized that, rather than grasping the full implication of Fleck's message during his early reading (relying on his "rusty German" as he put it), he had focused primarily on ". . . changes in the gestalts in which nature presented itself, and the resulting difficulties in rendering 'fact' independent of 'point of view.'"

While writing our book, Patrick Wells and I did not know of Fleck's perceptive analysis of scientific process, but our thoughts nevertheless had converged with his on many issues, particularly in his sections on epistemology.

Realism and Relativism Schools of Thought

Realism. The dubious notion that one can "know" reality was challenged repeatedly in Fleck's treatise. He also recognized that scientists become too committed to hypotheses. Fleck (1935/1979) wrote (p. 84): "Observation and experiment are subject to a very popular myth. The knower is seen as a kind of conquerer, like Julius Caesar winning his battles according to the formula 'I came, I saw, I conquered.'" And (p. 84): "Even research workers who have won many a scientific battle may believe this naive story when looking at their own work in retrospect." Later Fleck commented (p. 125): ". . . the [generated] fact becomes incarnated as an immediately perceptible object of reality."

The notion that "fact" has not necessarily been gained emerges from Fleck's statement (p. 32): "The liveliest stage of tenacity in systems of opinion is creative fiction, constituting, as it were, the magical realization of ideas and the interpretation that individual expectations in science are actually fulfilled."

Fleck's awareness of the essence of Duclaux's statement (above) is evident in his own statements (p. 27): "Once a structurally complete and closed system of opinions consisting of many details and relations has been formed, it offers enduring resistance to anything that contradicts it," and (pp. 30, 31): "The very persistence with which observations contradicting a view are 'explained' and smoothed over by conciliators is most instructive. Such effort demonstrates that the aim is logical conformity within a system at any cost . . ." Neither was Fleck blind to social constraints in the conduct of research (p. 47): ". . . . Social consolidation functions actively even in science. This is seen particularly clearly in the resistance which as a rule is encountered by new directions of thought."

One of the more striking features of Fleck's book is the notion of "thought collectives" (expanded upon below). Various interest groups exist within each scientific community, as exemplified today by units within the electronic mail system. He defined his use of the term "thought collective" as (p. 39): "a community of persons mutually exchanging ideas or maintaining intellectual interaction. . . ."

Any one person belongs to several thought collectives (very obvious in multiple enrollment in e-mail networks) and becomes molded into the thought patterns expected within each scientific community. Kuhn's notion of "paradigm hold" was already known to Fleck (p. 28): "When a conception permeates a thought collective strongly enough, so that it penetrates as far as everyday life and idiom and has become a viewpoint in the literal sense of the word, any contradiction appears unthinkable and unimaginable."

Relativism. We chose the term "relativism" for emphasis in our book, among other possible choices of words, to stress the relative nature of knowledge (see Excursus RE in Wenner & Wells, 1990), but Fleck had already perceived the same concept when he wrote (p. 50): "An empirical fact . . . is relative. . . . Both thinking and facts are changeable. . . . Conversely, fundamentally new facts can be discovered only through new thinking." In stronger words he wrote (p. 20): ". . . we would argue that there is probably no such thing as complete error or complete truth" and (p. 48): ". . . nobody has either a feeling for, or knowledge of, what physically is possible or impossible."

Fleck again recognized the influence of social factors in science (p. 124): "If a fact is taken to mean something fixed and proven, it exists only in vademecum science," and (p. 21): "At least three-quarters if not the entire content of science is conditioned by the history of ideas, psychology, and the sociology of ideas and is thus explicable in these terms."

Finally, Fleck recognized the very tentative nature of scientific investigation (pp. 10, 11):

The acquisition of physical and psychological skills, the amassing of a certain number of observations and experiments, the ability to mold concepts, however, introduce all kinds of factors that cannot be regulated by formal logic. Indeed, such interactions . . . prohibit any systematic treatment of the cognitive process.

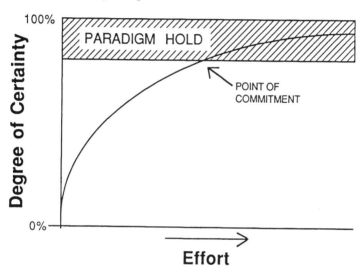

Figure 2. A diagram illustrating how individuals who rely too heavily on verification or falsification can become locked into a paradigm hold. Verificationists may accumulate support for a hypothesis but fail to test it. Those who attempt to falsify a null hypothesis may ignore anomalies and erroneously conclude that failure to falsify leads to truth. In either case , basic assumptions may then no longer be questioned.

Four Scientific Approaches: Fleck's Comments

In our book, Patrick Wells and I recognized that scientists use one or more of four approaches in the collective scientific process (as described above). Fleck had clearly preempted us on this score, more so on the first two than on the last two.

Exploration. Atkinson (1985) coined the term, "creation of an image," to describe what happens when one has a new perception about existing information (lower right hand corner in Figure 1). However, he was clearly preempted by Fleck, as is evident by Fleck's use of the word "genesis" in the title of his book, *Genesis and Development of a Scientific Fact*.

Fleck recognized the creative spark (p. 48): ". . . the ability to perceive scientifically is only slowly acquired and learned. Its prime manifestation is discovery. This occurs in a complex, socially conditioned way. . . ." He also stressed that the creative individual should recognize the strictures of "reality" (p. 30): "Discovery is . . . inextricably interwoven with what is known as error. To recognize a certain relation, many another relation must be misunderstood, denied, or overlooked." Fleck further noted the influence of social relationships (p. 123): ". . . the true creator of a new idea is not an individual but the thought collective . . . The collective remodeling of an idea has the effect that, after the change in thought style, the earlier problem is no longer completely comprehensible."

If a novel idea does gain sufficient acceptance in the scientific community (rather readily if exotic, it seems), a new field of research may emerge, and many of those in "vademecum science" fall into line behind those in the "vanguard," in Fleck's terms. Soon assumptions become accepted as "facts" and become the basis for more "normal science" (Kuhn's term).

Verification. Perhaps one of the most rigid notions (and roadblocks) in animal behavior studies is the concept that one can prove something true if one gathers sufficient "positive" evidence (e.g., Wilson, 1972, p. 6). Fleck saw through that line of reasoning and recognized the additional social element involved (p. 37): ". . . once a statement is published it constitutes part of the social forces which form concepts and create habits of thought."

Garrett Hardin (1993, p. 225) expressed this same thought somewhat differently: "Often the creation of a noun ('substantive') seems to presume the presence of a substance, a physical *thing*." As an example, we have seen in behavioral studies that various expressions (e.g., "innate releasing mechanism" and "fixed action potential") come into vogue for a time and then disappear. Schneirla's school perceived that same problem: the invention of such terms brings us no closer to understanding (e.g., Tobach & Aronson, 1970, p. xvi): "He was especially opposed to such ethological terms as innate releasing mechanism, vacuum and displacement reactions, fixed action potential, and action specific energy, which he considered to be reifications deduced from the basic assumption of the existence of instincts."

The problem here is that members of the scientific community too readily accept notions that may not be backed by substantial evidence, as illustrated by Mark Twain's comment in *Life on the Mississippi*: "There is something fascinating about science. One gets such wholesale returns of conjecture out of such a trifling investment of fact."

In Fleck's words (p. 42): "Thoughts pass from one individual to another, each time a little transformed. . . . Whose thought is it that continues to circulate? It is one that obviously belongs not to any single individual but to the collective," and (p. 28): "When a conception permeates a thought collective strongly enough . . . , any contradiction appears unthinkable and unimaginable."

Eventually, the "thought collective" forms a consensus and holds to given positions, as outlined by Fleck (p. 27):

1. *A contradiction to the system appears unthinkable.*
2. *What does not fit into the system remains unseen;*
3. *alternatively, if it is noticed, either it is kept secret, or*
4. *laborious efforts are made to explain an exception in terms that do not contradict the system.*
5. *Despite the legitimate claims of contradictory views, one tends to see, describe, or even illustrate those circumstances that corroborate current views and thereby give them substance.*
In other words, "vademecum science."

Falsification. Fleck clearly understood the importance of heeding evidence that did not fit prevailing thought. He clearly preempted Kuhn on the notion of "paradigm hold" when he wrote (p. 27): "Once a structurally complete and closed system of opinions consisting of many details and relations has been formed, it offers enduring resistance to anything that contradicts it."

In the past few decades of animal behavior studies we have seen numerous examples of some rather exotic hypotheses being put forth by vanguard scientists and embraced by vademecum science participants. As examples we have had honey bee, dolphin and chimpanzee "languages." In the "hard sciences" examples include "cold fusion," "water with a memory," and "polywater" (e.g., Rousseau, 1992).

Fleck's attitude about scientific process is relevant to the fact that progress in animal behavior and ecology studies seems to be slow (e.g., Dayton, 1979). One might also find relevant statements about scientific protocol in Garrett Hardin's book, *Living Within Limits: Ecology, Economics, and Population Taboos* (1993). Hardin (p. 41) wrote that if "each new proposal advanced" were to be "assumed to be true until it is proven false . . . the scientific community would soon be overwhelmed by unworkable proposals, and the advance of science would be greatly retarded." But many do not seem to realize that something not yet proven false is not necessarily true.

Inference. There seems to be little in Fleck's book that relates directly to Chamberlin's concept of the multiple inference approach, but Fleck had a good grasp of the relative nature of science when he wrote (p. 20): ". . . we would argue that there is probably no such thing as complete error or complete truth." One can only speculate about how Fleck might have benefited from Chamberlin's thoughts (1890/1965), as resurrected by Platt (1964).

As Eugene Meyer phrased it (personal communication), Fleck was too "gloomy" on this point. The multiple inference approach has great promise for animal behavior studies, once that procedure becomes part of the research arsenal for those in that field. Whereas "truth" may always be elusive, diverse options can always be kept alive with this approach; furthermore, paradigm holds will likely be less severe. In our work the multiple inference/strong inference approach has been extremely valuable (e.g., Wenner et al., 1969; Wenner, 1972; Wenner & Harris, 1993).

Thought Collectives

For quite some time, individuals within the "hard sciences" have insisted that social factors do not influence the conduct of their science. In the Foreword to Fleck's book, Kuhn wrote (1979, p. viii): ". . . in 1950 and for some years thereafter I knew of no one else who saw in the history of science what I was myself finding there . . . acquaintance with Fleck's text helped me to realize that the problems which concerned me had a fundamentally sociological dimension." More recently, Hardin (1993, p. 257) echoed that thought: "Part of the unofficial mythos that supports science is the belief that truth will prevail, no matter what. If you have a heretical idea, publish it, supporting it with data and arguments as needed, it will be noticed. If your theory is true it will soon be accepted by the establishment; heterodoxy will metamorphose into orthodoxy."

Fleck's use of the term "thought collective" can now be seen to be especially apt. Increasingly, those who study the sociology and psychology of science recognize the importance of social bonds among those who work within given broad research groups (in Fleck's words, again, "thought collectives"). He lucidly described the social hierarchy we now recognize (p. 124): "Every discipline . . . has its own *vanguard*. . . . This is followed by the *main* body, the official community [*vademecum* scientists]. Then come the somewhat disorganized stragglers."

Fleck elaborated upon that point (p. 41): "The individual within the collective is never, or hardly ever, conscious of the prevailing thought style, which almost always exerts an absolutely compulsive force upon his thinking and with which it is not possible to be at variance," and (p. 82): "The more deeply one enters into a scientific field, the stronger will be the bond with the thought collective and the closer the contact with the scientist." In 1841 Mackay was more pointed (in Taubes, 1993, p. 107): "Men, it has been well said, think in herds; it will be seen that they go mad in herds, while they only recover their senses slowly, and one by one."

The Schneirla group had a similar perception of the problem (e.g., Lehrman, 1970, p. 21): "It is too easy to close one's mind to an argument by simply deciding that the source of the argument is an outsider."

Fleck used even more powerful language when he wrote (p. 141):

To the unsophisticated research worker limited by his own thought style, any alien thought style appears like a free flight of fancy, because he can see only that which is active and almost arbitrary about it. His own thought style, in contrast, . . . becomes natural and, like breathing, al-

most unconscious, as a result of education and training as well as through his participation in the communication of thoughts within his collective.

Hardin punctuated that idea in other words (1993, p. 43): "The scientific mind is not closed: it is merely well guarded by a conscientious and seldom sleeping gatekeeper. . . ."

Consider now the situation in which two thought collectives embrace different concepts of "reality" on a given topic. It may become impossible for them to communicate, one with another. Controversy can then erupt. Kuhn addressed that point in his foreword to Fleck's book (p. ix): ". . . given my own special concerns, I am particularly excited by Fleck's remarks on the difficulties of transmitting ideas between two 'thought collectives,' above all by the closing paragraph on the possibilities and limitations of participation in several 'thought communities.'"

From that point one can see how someone in a thought collective who attends too closely to arguments of an opposing thought collective might go beyond the bounds of acceptable behavior (see Pauly, 1981, with reference to Jacques Loeb). We can see here a gradation along a continuum—loyal club member to skeptic (pessimist). Only a little step more, and one can fall into the "whistle-blower" category (the hypothesis fails too often). Hardin (p. 234) addressed that point: "Cronyism can be good, cronyism can be bad. 'Whistle-blowers,' who seek to serve the good of a larger group . . . [science or society as a whole instead of merely the thought collective] . . . are, more often than not, ostracized by their fellow workers." Glazer and Glazer (1986) covered that theme in some detail.

Thinking

Fleck became more profound when he considered the role of "thinking" in scientific research. He suggested that individuals were not normally capable of independent thinking and used a quotation from Gumplowicz to emphasize that point (p. 46): "The greatest error of individualistic psychology is the assumption that a *person* thinks . . ."; he further commented (p. 47): "What actually thinks within a person is not the individual himself but his social community," and (p.

98): ". . . thinking [is] a supremely social activity which cannot by any means be completely localized within the confines of the individual."

Fleck realized that he had been too negative on that point and later qualified that rather rigid stand by using a quotation from Jerusalem (p. 49): "Man acquires [the ability to think 'purely theoretically' and to state 'given facts purely objectively'] only slowly and by degrees, to the extent that by conscious effort he overcomes the state of complete social bondage and thus develops into an *independent* and *self-reliant* personality." Fleck thus felt it possible that individuals could become aware of themselves and of the role of social forces in the scientific process.

Reflections on Scientific Controversy

With the above comments by Fleck as background, we can consider further the role of controversy in scientific research, with some special attention to the question of "language" in honey bees. Although this section is written by an insider, it has the unique perspective of one who stepped out of a controversy for more than two decades.

The two-decade leave of absence mattered little. Even though my colleagues and I were not directly involved, the controversy continued unabated during those two decades. Honey bee "language" proponents repeatedly felt compelled to conduct "the definitive experiment," one that could reinforce the prevailing viewpoint ("consensus") of honey bee "language." In a series of critiques spanning that same period, Rosin (e.g., 1978, 1980, 1988, 1992) exposed the flaws in those "definitive" experiments in terms of theory, design and execution, critiques that relied heavily on theoretical foundations laid by Maier, Morgan and Schneirla. For example, Rosin wrote (1988, p. 268):

Whereas Wenner & Wells (1987) explain that they came to oppose the "dance language" hypothesis as a direct result of their own research in honey bee behavior, I joined their opposition due to no particular interest in honey bees, but because I saw in the specific "dance language" controversy a major reflection of a much more generalized and much more important controversy over the whole field of animal behavior, between European Ethology . . . and Schneirla's School. . . .

Note the close correspondence here between Rosin's comment and Gerard Piel's comment in the first paragraph of this contribution.

Rosin earlier had also written (1978, p. 589): "The controversy between the von Frisch group's 'language' hypothesis . . . and Wenner's group's olfactory hypothesis . . . for the arrival of honey bee recruits at field sources, is essentially a controversy between a human-level hypothesis for an insect and an insect-level hypothesis for an insect."

Even though Rosin's arguments were largely informally dismissed by the bee language "thought collective" (essentially by near lack of citation), that same community eventually recognized that each of the various "definitive" experiments had been inadequate. Latour (1987, p. 43) addressed this type of issue: "If an article claims to finish the dispute once and for all it might be immediately dismembered, quoted for completely different reasons, *adding* one more empty claim to the turmoil" (Latour's emphasis).

Fleck's contribution about "thought collectives" helps place this last point in perspective. Just as it is not usually the individual but the thought collective that is creative, neither are controversies merely between individuals. They are instead most often controversies between thought collectives. Even though various sociopolitical maneuvers can result in denial of a platform to particular individuals, the thought collective to which those individuals belong doesn't simply vanish. All that happens is that one or more key figures in one thought collective may lose a platform and become (hopefully, to those in an opposing thought collective) temporarily silent. Whereas members of the opposing thought collective can then convince themselves that an issue has been finally resolved in their favor, members of the original thought collective can continue to challenge any "elegant" experiments that form the basis for continued consensus.

Thus, it is usually not two individuals who are engaged in a nontrivial controversy. Rather, it is two thought collectives with two different views of "reality" that collide. Either that, or one thought collective with a fixed view of "reality" collides with another thought collective that may recognize that scientific accomplishments always remain relative.

To ascertain the degree to which the thinking of an individual may be unconsciously controlled by expectations of one or more thought collectives, one need only ask a generic question: Is it conceivable that your assumption (hypothesis) is not true? The answer to that question reveals much—an unquestioning member of a thought collective usually immediately answers in the negative ("no, that is not possible"). As Fleck put it (p. 107): "At a certain stage of development the habits and standards of thought will be felt to be the natural and the only possible ones. No further thinking about them is even possible."

Thomas Kuhn termed the above fixation, "paradigm hold" (see Figure 2), but Fleck had earlier appreciated the same concept when he wrote (as quoted earlier) (p. 27): "Once a structurally complete and closed system of opinions consisting of many details and relations has been formed, it offers enduring resistance to anything that contradicts it." Hardin (1993, p. 4) used another interesting expression, "gatekeeper of the mind," for much the same idea but also recognized just how insidious such fixation can become (p. 4): "An effective gatekeeper of the mind does not call attention to itself. It actuates a psychological mechanism called a *taboo*."

Dewsbury (1993, p. 869) attempted to justify some censorship of novel perspective, viewing that practice as a necessary evil:

Both the creative innovator and the crackpot work at the fringes of the prevailing paradigm, and it often is difficult to distinguish one from the other in the early stages of development. The scientific establishment, therefore, must develop a commitment to scientific orthodoxy that makes it hostile to challenges to that orthodoxy. Limiting access to the publication outlets controlled by the scientific establishment is one way in which those who are part of a scientific in-group or who are working within the dominant perspective can help defend that perspective.

If challenges to established dogma are not permitted, however, science does not advance.

Consider now the attitudes of scientists toward controversy. If controversy erupts in some field of science other than our own, we can enjoy watching the antics of those committed to the dogma (and/or flawed protocol) of one thought collective or another. Biologists and psychologists, for example, may well relish discussions about controversies in the "hard sciences" mentioned earlier.

Consider further the concept of taboo (e.g., Hardin, 1993). If a nontrivial controversy is too close to home, "vademecum scientists" in one thought collective or another distance themselves from the emerging controversy. Those who become enmeshed in the controversy become suspect (have only themselves to blame—a self-destructive act) unless they are a part of "vanguard science," those "elite" thought collective members who spearhead goals of their own thought collective. Eventually no real interchange occurs between or among members of opposing thought collectives. Such extracommunity interaction is part of the taboo. As Hardin saw it (p. 4): "Westerners, with their cherished tradition of free speech and open discussion . . . change the subject."

Here we must differentiate between minor and major controversies. Members of a thought collective may good-heartedly engage in minor controversies ("playful" or trivial controversies) as long as they remain minor; that is, when none of the basic assumptions of the thought collective become threatened. Participants may even pride themselves in their tolerance of divergent opinions.

Different fields have different levels of tolerance for revolutionary ideas, as is evident in the speed with which controversies become resolved. Fast-moving fields (e.g., genetics, molecular biology, nuclear physics), ones that routinely employ the strong inference approach in research (e.g., Platt, 1964), are generally much more receptive to challenge of existing dogma—which is why they are fast moving. Even in fast-moving fields, though, true progress can sometimes be slow, as in the rate of adoption of the chemiosmotic coupling hypothesis proposed by Peter Mitchell (Gilbert & Mulkay, 1984).

Once a major controversy erupts, certain events are quite predictable. Resolve within each thought collective solidifies, and much private support is given to those in the front lines ("vanguard scientists"). Papers submitted for publication by vanguard scientists do not undergo the same scrutiny as those submitted by vademecum scientists or those outside the thought collective, a point thoroughly documented by Peters and Ceci (1982, as summarized in Wenner & Wells, 1990, p. 191). Neither does totality of evidence count for much. Participants select those bodies of evidence that reinforce their own position. An example of such resolve was documented by Taubes

(1993, p. 270): "Cold fusion existed until proven otherwise. . . . The Electrochemical Society administrators wanted to avoid a repetition of the rampant negativity of the Baltimore American Physical Society meeting. Speakers would present 'confirmation results' only."

Those in the Schneirla school understood this type of development, as concisely stated by Lehrman (1970, pp. 18, 19):

When opposing groups of intelligent, highly educated, competent scientists continue over many years to disagree, and even to wrangle bitterly about an issue which they regard as important, it must sooner or later become obvious that the disagreement is not a factual one, and that it cannot be resolved by calling to the attention of the members of one group (or even of the other!) the existence of new data which will make them see the light. Further, it becomes increasingly obvious that there are no possible crucial experiments that would cause one group of antagonists to abandon their point of view in favor of that of the other group.

My colleagues and I encountered that phenomenon after we tested the honey bee dance language hypothesis and found it wanting. In the following two decades, symposia on insect communication included only participants who could provide positive results in support of "language" among bees, in spite of an early comment (Wells & Wenner, 1973, p. 175):

Do honey bees have a language? That is a question which may never be answered with certainty. It may be more useful to examine assumptions critically, state hypotheses and their consequences with precision, review the evidence objectively, and ask: Can we now believe that honey bees have a language? Thus, it appears that the honey bee forager recruitment controversy is not about the nature of evidence but rather about the nature of hypotheses. It is not what investigators observe (the data) but what they believe (infer) that is at the heart of the controversy.

When one realizes the importance of thought collective control, it becomes more clear why journal referees come down strongly on one side or another during a controversy, just as do proposal reviewers, members of panels for granting agencies and even members of the media (e.g., Horgan, 1990, p. 29; Taubes, 1993, p. 263).

That is why new scientific journals have often been started after members of one thought collective have been excluded from existing platforms by members of other thought collectives (see Hull, 1988).

Fleck again preempted us all when he addressed that aspect of scientific controversy (p. 43): "Words which formerly were simple terms become slogans; sentences which once were simple statements become calls to battle." Participants in a given controversy, it seems, suddenly fail to distinguish between ideas and personalities. Daniel Lehrman, of the Schneirla school, recognized this type of development (1970, p. 47): "We do not lightly give up ideas which seem central to us, and when they are attacked, we tend to mobilize defenses against the attacks."

When controversies mature, positions almost imperceptively change, a point clearly emphasized by some of those in the Schneirla school (e.g., Lehrman, 1970, p. 47). We can restate

The attacked ideas in such a form as to make them seem again convincing to an audience whose confidence in them might have been weakened by the criticism. But when we change the formulation of the ideas in such a situation, we may also be modifying the ideas themselves, in response to criticisms which really may have been leveled against weaknesses in the original formulations.

In time, modifications can have accumulated to such an extent that the original hypothesis becomes lost. This event seems to have happened with respect to the dance language hypothesis, leading us to comment (Wenner et al., 1991, p. 771): ". . . proponents of the dance language hypothesis today no longer seem to have a clear notion of what one should expect from that hypothesis."

Controversies eventually become resolved, if not automatically or rapidly, by renewed attention to Nature, which cannot be fooled by rhetoric. In many cases, if not most, that resolution comes *gradually* (as in Lehrman's comment, above). One of the prevailing hypotheses becomes obviously more useful than the other(s) for explaining *all* available evidence (e.g., Wenner & Wells, 1987). But no prizes will be forthcoming: Fleck wrote (p. 123): ". . . the collective remodeling of an idea has the effect that, after the change in thought style, the earlier problem is no longer completely comprehensible." Hardin (p. 51) echoed that thought: "Only at the end of an era do surviving pessimists have a chance to be recognized by their fellow citizens as being (finally) right, but it is not likely that they will then be praised for their foresight."

Neither do we find that textbooks or the popular literature treat the resolution of controversy adequately. Textbook writers (usually "vademecum" scientists, or "stragglers" in Fleck's terms) and editors continue to select items from earlier texts without realizing that some "facts" have become discredited (e.g., Paul, 1987). Even in the late 1980s I found a clip in the *Santa Ynez Valley News* (California) describing how some flies could travel 880 mph, despite discreditation of that "fact" in 1938 (see Wenner, 1989).

Today we realize that paradigm holds are essential in science; otherwise it would not be possible to design and conduct experiments. Nevertheless, we must make ourselves and others in our thought collectives more fully aware that paradigms can control our thinking and that we need to always reexamine assumptions (see *Ten Principles*, below).

The Complicated World of Animal Behavior Studies

The study of animal behavior is actually one of the more difficult tasks in science, occasioned in large part by the twin pitfalls of teleology (e.g., Bernatowicz, 1958) and anthropomorphism. The rationale among all too many participants: If a behavior is there, it must be good for something. Furthermore, its function must correspond with our current concept of reality (i.e., it has to be good for what we human beings think it should be good for).

The first of those pitfalls (functional approach) has evolved out of our Judeo-Christian heritage (God created all for a given purpose). In the fields of animal behavior and ecology, in recent years this attitude has shifted to a belief in "Nature's purpose." (See Excursus TEL in Wenner & Wells, 1990 for a more expanded treatment of this topic.)

The second pitfall (assigning human characteristics to our subjects) has been inculcated in all of us since childhood. The "Nature" programs on television promulgate that concept throughout society (the "Disneyfication of science" as zoologist Bill Tavolga once phrased it).

Schneirla was a leader in opposing those twin pitfalls (e.g., Tobach & Aronson, 1970, p. xvi): "He was guided overall by the law of parsimony, by Morgan's canon, and above all, by the need to avoid the dangers and fallacies of anthropomorphism and zoomorphism." Instead, Schneirla advocated use of the inductive method in animal behavior research. Unfortunately, we have recently been treated to a resurgence in the "animal thinking" concept (including "cognition") in nonhuman species (e.g., Griffin, 1984), a practice that Schneirla decried (as in Gerard Piel, 1970, p. 3):

One who sets out to demonstrate that protozoan organisms or any others have the mental characteristics of man may convince himself at least, provided he singles out opportunely the brief episodes which seem describable as instances of perception of danger, of reasoning, or what not. By the same method, the absence of reasoning in man can be proved with ease.

Also, one need not watch many Nature programs on television before it becomes evident that "exotic" episodes carry the day. This emphasis on the exotic carries with it an implicit pressure on students of animal behavior; experimental subjects and results must be exciting. Actually, so is it with all of scientific research.

Self and Mass Delusion

At times claims by those on the "forefront" of science may stretch the bounds of credibility (e.g., "cold fusion" research). Nevertheless, support may arise from varied quarters, and a vanguard scientist may then become ever more committed (as in Atkinson, 1985) to an exotic hypothesis. Optimism rules, in science as in other aspects of life, as expressed in another context by Hardin (1993, p. 50): "Perhaps for several decades the optimist will win out—getting richer, earning more prestige in the community, marrying better, and perhaps having more children than the pessimist . . . thus is the pessimist made to look foolish in the short run."

The Nobel chemist, Irving Langmuir (in unpublished lecture notes, Taubes, 1993, pp. 342, 343) was apparently the first to apply the label, "pathological science," to circumstances such as "supersonic flies" (e.g., Wenner, 1989), "polywater" (e.g., Rousseau, 1992), and "cold fusion" (e.g., Taubes, 1993). Languir defined pathological science in six points, as follows:

1. *The maximum effect that is observed is produced by a causative agent of barely detectable intensity, and the magnitude of the effect is substantially independent of the intensity of the cause.*
2. *The effect is of a magnitude that remains close to the limit of acceptability.*
3. *Claims of great accuracy.*
4. *Fantastic theories contrary to experience.*
5. *Criticisms are met by* ad hoc *excuses thought up on the spur of the moment.*
6. *Ratio of supporters to critics rises up to somewhere near 50 percent and then falls gradually to oblivion.*

Langmuir added a comment: "The critics can't reproduce the effects. Only the supporters could do that. In the end, nothing was salvaged. Why would there be? There isn't anything there. There never was." However, one can see, even here, a remarkable parallel between Langmuir's comments and those of Fleck earlier in 1935, as outlined above in Fleck's emphasis on the role of "consensus" within thought collectives.

Of course, there comes a time when the hard face of reality peers down, and the bubble may break. Three decades ago we were treated to supposed examples of learning by means of cannibalism. Likewise, the honey bee "dance language" hypothesis was with us for 20 years before it was truly tested by experiment, and Moore (1988) has now finally raised unsettling issues about the presumed "magnetic compass" orientation of pigeons, three decades after that "fact" was established. Who knows which of the other leading animal behavior hypotheses might fail critical tests once animal behavior studies evolve to the point where careful scrutiny of both evidence and basic assumptions becomes a part of the working protocol?

Unfortunately, the exotic sells (as indicated above), and those in vanguard science are not immune from self-delusion (the easiest one to fool is oneself). Fleck had an appropriate comment about that circumstance as well (p. 105): "The elite panders, as it were, to public opinion and strives to preserve the confidence of the masses. This is the situation in which the thought collective of science usually finds itself today [in 1935]. If the elite enjoys the stronger position, it will endeavor to maintain distance and to isolate itself from the crowd."

Promotion in the Media
Scientists in today's world race to the media with their most recent "finds," as is evident from the activity present in the press rooms established at major scientific conventions. Furthermore, science reporters no longer "report" and evaluate the news they have been exposed to. They seek advice from "authorities" in the field and search for "consensus" among those experts, seemingly afraid to think for themselves. Little do they realize that their actions may well do no more than reinforce the views of one particular "thought collective" (as phrased by Fleck) over another, justified or not.

The curious notion of "polywater" is a good example. An obscure physicist invented the polywater hypothesis, an exotic notion with military implications that caught on all too rapidly and widely (e.g., Franks, 1981; Wenner & Wells, 1990, pp. 49–50; Rousseau, 1992). After a decade of intensive research devoted to verification of the hypothesis by many on several continents, accompanied by much media hype, Denis Rousseau (e.g., 1992) conducted a critical experiment, the results of which exposed earlier findings as artifacts. For a time members of the polywater thought collective were upset, and his results were ignored, but eventually Nature won out. "Cold fusion" ran much the same course in only a few months (Taubes, 1993).

Rousseau's experience, coupled with more recent episodes of "water with a memory" and "cold fusion," led him to the conclusion that all these events constitute "pathological science" in Langmuir's sense, events that can be (and should have been) recognized early on—but only if one is aware of three characteristics that they have in common (Rousseau, 1992). The first two were as in points 1 and 2, as well as 4, of Langmuir (above), but Rousseau (1992, p. 54) added a third: "To avoid these pitfalls, scientists must conceive and carry out a critical series of experiments. . . . But the third identifying trait of pathological science is that the investigator finds it nearly impossible to do such experiments."

I might add to that third point the fact that scientists locked into a paradigm simply can neither recognize nor accept the results of tests or critical experiments that counter the belief system of their thought collective. Neither can they bring themselves to repeat such "negative" experiments. All too often in animal behavior studies "critical experiments" have come to be identified only as those that support a prevailing hypothesis. Unfortunately, the media can keep a discredited hypothesis alive long after it is no longer useful or believed by the scientists themselves. An example: How long will honey bee "dance language" persist in school texts, magazine articles, video tapes, Nature programs and newspaper clips if (or after) the scientific community abandons that line of research?

One could add yet another point: The present unfortunate media and public focus on *individual accomplishment* in science rather than an emphasis on the *collective nature* of the scientific process. Prizes and awards seem to go most often to those who reinforce the belief system of one thought collective or another, rather than to those who may truly advance science by exploding a myth, as Rousseau did. Hardin addressed that point (p. 109): "Today the greatest honor is accorded to speakers who focus on individual interests to the exclusion of community interests."

Ten Principles of Scientific Research

The bee . . . extracts matter from the flowers of the garden and the field, but works and fashions it by its own efforts. The true labor of philosophy resembles hers, for it neither relies entirely nor principally on the powers of the mind, nor yet lays up in the memory the matter afforded by the experiments of natural history and mechanics in its raw state, but changes and works it in the understanding. (Francis Bacon 1620/1952, p. 126)

We can look more deeply into history for guidelines that we can follow, and then instill in our students the importance and excitement of *process*, rather than the *accomplishments* of science. Throughout history any given person may have had a good grasp of at least a portion of the scientific process (Figure 1). Which of the portions was grasped, however, has varied considerably among individuals and reminds one of different views of the elephant held by several blind men, each of whom had touched a different part of its body (e.g., Atkinson, 1985).

A common misconception among scientists is that "objectivity" is possible—they forget that we are human beings first and scientists second. When I mentioned to a young faculty member, new at the University of California (not our campus), that his mentor in his former graduate program might have had a strong bias on one point, he countered: "No" and said that his former advisor "is totally objective and without bias." Furthermore, the temptation to exaggerate findings, fudge results, omit unfavorable data, emphasize "positive" results, etc., may have become even greater under "big" science than it was before, especially under current intense pressures to publish and/or to procure grants.

We have to recognize that none of us can be truly objective. Regrettably, that new faculty member had not learned in his university education that each of us is biased to a greater or lesser degree on a great many issues. As scientists, we are no exception (e.g., Mahoney, 1976). Scientists and observers of science (e.g., sociologists, psychologists and philosophers), in fact, have spent considerable time wondering how it is that scientists succeed, given the very human nature of us all.

In 1991, students in a class I conducted ("The Nature of Biological Science") in the College of Creative Studies on my home campus surveyed the literature and collectively formed a list of principles that scientists should be aware of as they conduct their research. Those points are presented below, accompanied by appropriate quotations and citations, in chronological order within each section.

1. **Attend more to Nature than to theory**
    - Aristotle (330 B.C./1931, III, p. 760,b): "... we should trust more the observations than the theory, and we should hold good the latter only if facts support it."
    - Francis Bacon (1620/1952, p. 84): "... it is the greatest weakness to attribute infinite credit to particular authors ... They who have presumed to dogmatize on nature ... have inflicted the greatest injury on philosophy and learning. For they have tended to stifle and interrupt inquiry."

- Louis Pasteur (in Rene Dubos, 1950, p. 376): Preconceived ideas "become a danger only if [an experimenter] transforms them into fixed ideas . . . The greatest derangement of the mind is to believe in something because one wishes it to be so."
- Claude Bernard (1865/1957, p. 39): "In a word, we must alter theory to adapt it to nature, but not nature to adapt it to theory. . . . When we meet a fact which contradicts a prevailing theory, we must accept the fact and abandon the theory, even when the theory is supported by great names and generally accepted."
- Evelyn Fox Keller (1983, p. 35): ". . . the necessary next step seems to be the re-incorporation of the naturalist's approach—an approach that does not press nature with leading questions but dwells patiently in the variety and complexity of organisms."
- Naomi Aronson (1986, p. 630): "The production of scientific knowledge is simultaneously the production of scientific error. . . ."
- David Bohm and David Peat (1987, p. 51): ". . . to cling rigidly to familiar ideas is in essence the same as blocking the mind from engaging in creative free play."

2. **Use the appropriate scientific approach**

- Claude Bernard (1865/1957, p. 34): "The experimental method . . . cannot give new and fruitful ideas to men who have none; it can serve only to guide the ideas of men who have them, to direct their ideas and to develop them as to get the best possible results."
- Louis Pasteur (in Emile Duclaux, 1896/1920, p. 97): "Repeat [the experiments] with the details which I give you and you will succeed just as I have done."
- Thomas Chrowder Chamberlin (1890/1965, p. 756): "The effort is to bring up into view every rational explanation of new phenomena, and to develop every tenable hypothesis respecting their cause and history. The investigator thus becomes the parent of a family of hypotheses and, by his parental relation to all, he is forbidden to fasten his affections unduly upon any one."
- John R. Platt (1964, p. 350): "When multiple hypotheses become coupled to strong inference, the scientific search be-

comes an emotional powerhouse as well as an intellectual one."

• Max Silvernale (1965, p. 4): "There is nothing really mysterious about the scientific method; it is so simple that it can be understood by almost everyone. Yet the sad truth is that the majority of people today are not scientific, even though ours is a scientific age."

• Belver Griffith and Nicholas Mullins (1972, p. 963): "In our examination of highly coherent groups, we see two factors [in addition to communication] as basic to science: first, the radical revision of scientific theory and method . . .; second the rarity of high levels of personal creativity."

• Bernard Dixon (1973, p. 34): ". . . the top-class creative thinker designs a single, crucial experiment that decides absolutely between one hypothesis and another."

• David Bohm and David Peat (1987, p. 100): ". . . science should be carried out in the manner of a creative dialogue in which several points of view can coexist, for a time, with equal intensity."

3. **Seek understanding, not "truth"**

• Claude Bernard (1865/1957, p. 23): ". . . (An experimenter) must submit his idea to nature and be ready to abandon, to alter or to supplant it, in accordance with what he learns from observing the phenomena which he has induced."

• Peter Medawar (in Bernard Dixon, 1973, p. 24): "Truth takes shape in the mind of the observer; it is his imaginative grasp of *what might be true* that provides the incentive for finding out, so far as he can, what *is* true" (emphasis Medawar's).

• Alan Chalmers (1982, p. xvi): "There is just no method that enables scientific theories to be proven true or even probably true."

• James Atkinson (1985, p. 734): Science is "a process whereby the human capacity for imagination creates and manipulates images in the mind. . . . This process produces "concepts, theories, and ideas which incorporate and tie together shared human sensory experience and which are assimilated into human culture through a similar act of re-creation."

• Richard Feynman (in Gleick, 1992, p. 438): "I don't have to know an answer. I don't feel frightened by not knowing things, by being lost in a mysterious universe without any purpose, which is the way it really is as far as I can tell."

4. **Recognize limits of human perception**

• Claude Bernard (1865/1957, p. 23): "It is impossible to devise an experiment without a preconceived idea." (1865/1957, p. 38): "[Those] who have excessive faith in their theories or ideas are not only ill prepared for making discoveries; they also make very poor observations"; (1865/1957, p. 52): "The doubter is a true man of science; he doubts only himself and his interpretations, but he believes in science."

• Jerome Bruner and Leo Postman (1949, p. 222): "When . . . expectations are violated by [Nature], the perceiver's behavior [is one of] resistance to the recognition of the unexpected or incongruous."

• Evelyn Fox Keller (1983, p. 145): "In practice, scientists combine the rules of scientific methodology with a generous admixture of intuition, aesthetics, and philosophical commitment."

• Lewis M. Branscomb (1985, p. 422): "Nature does not 'know' what experiment a scientist is trying to do. 'God loves the noise as much as the signal.'"

5. **Be honest and accurate (careful)**

• Peter Medawar (1979, p. 39): "A scientist who habitually deceives himself is well on the way toward deceiving others."

• Lewis M. Branscomb (1985, p. 423): ". . . integrity is essential for the realization of the joy that exploring the world of science should bring to each of us."

• Walter Stewart and Ned Feder (1987, p. 214): "Scientists have to an unusual degree been entrusted with the regulation of their own professional activities. Self-regulation is a privilege that must be exercised vigorously and wisely, or it may be lost."

• Efraim Racker (1989, p. 91): "The spiritual damage caused by scientific fraud is irreversible, and those involved are and should be reported and prosecuted irrespective of whether financial losses are involved."

6. **Pursue reasons for anomalies**

• Claude Bernard (1865/1957, p. 23): "An experimenter, who clings to his preconceived idea and notes the results of his experiment only from this point of view, falls inevitably into error, because he fails to note what he has not foreseen and so makes a partial observation." ". . . [the experimenter] must never answer for [nature] nor listen partially to her answers by taking from the results of an experiment, only those which support or confirm his hypothesis. We shall see later that this is one of the great stumbling blocks of the experimental method"; (1865, p. 50): "Experimenters . . . always doubt even their starting point"; (1865, p. 56): "Some . . . fear and avoid counterproof. As soon as they make observations in the direction of their ideas, they refuse to look for contradictory facts, for fear of seeing their hypothesis vanish."

• Thomas Kuhn (1962/1970, pp. 52, 53): "Discovery commences with the awareness of anomaly, i.e., with the recognition that nature has somehow violated the paradigm-induced expectations that govern normal science."

• Evelyn Fox Keller (1983, p. 123): "When scientists set out to understand a new principle or order, one of the first things they do is look for events that disturb that order. Almost invariably it is in the exception that they discover the rule"; (1983, p. 179): "The challenge for investigators in every field is to break free of the hidden constraints of their tacit assumptions, so that they can allow the results of their experiments to speak for themselves."

7. **Heed results of earlier workers**

• Louis Pasteur (in Rene Dubos, 1950, p. 376): "Preconceived ideas are like searchlights which illumine the path of the experimenter and serve him as a guide to interrogate nature.

• Belver Griffith and Nicholas Mullins (1972, p. 961): "[A] general indifference to the work of other researchers can generate considerable antagonism."

• Walter Stewart and Ned Feder (1987, p. 213): "Every scientist has, at a minimum, an obligation to ensure that what is published under his name is accurate."

8. **Focus on results, not on personalities**

• Belver Griffith and Nicholas Mullins (1972, p. 959): "Communication and some degree of voluntary association are intrinsic in science, and the important question therefore becomes not whether scientists organize, but rather how, why, and to what degree."

• Bernard Dixon (1973, p. 171): "Very, very few scientists appear to be aware that all experience . . . is subjective."

• Albert Rees (in the preface to the series [Medawar, 1979, p. xi]): Science "is an enterprise with its own rules and customs, but an understanding of that enterprise is accessible to any of us, for it is quintessentially human."

• William Broad and Nicholas Wade (1982, p. 180): "Like any other profession, science is ridden with clannishness and clubbiness. This would be in no way surprising, except that scientists deny it to be the case. . . . In fact, researchers tend to organize themselves into clusters of overlapping clubs."

9. Seek "how" and "what," not why

• Claude Bernard (1865/1957, p. 80): "The nature of our mind leads us to seek the essence or the *why* of things . . . experience soon teaches us that we cannot get beyond the *how*, i.e., beyond the immediate cause or the necessary conditions of phenomena" (emphasis Bernard's).

• John Steinbeck (1941/1962, p. 143): "But the greatest fallacy in, or rather the greatest objection to, teleological thinking is in connection with the emotional content, the belief. People get to believing and even to professing the apparent answers thus arrived at, suffering mental constrictions by emotionally closing their minds to any of the further and possibly opposite 'answers' which might otherwise be unearthed by honest effort."

10. Encourage expression of opposing views

• Peter Medawar (1979, p. 39): "*I cannot give any scientist of any age better advice than this; the intensity of the conviction that a hypothesis is true has no bearing on whether it is true or not*" (emphasis Medawar's).

• Bruno Latour (1987, p. 97): "As long as controversies are rife, Nature is never used as the final [arbitrator], since no one knows what she is and says. But *once the controversy is settled*, nature is the ultimate referee" (emphasis Latour's).

• Adrian Wenner and Patrick Wells (1990, p. 268): ". . . we have to get across the point at all levels that we are scientists because it is fun, and that the interjection of humor and tolerance of disparate viewpoints in our scientific controversies is a part of the fun, and is a part of life itself."

Perspective

A clear message emerges from the above. Each generation, unaware of the above time-tested principles (through lack of appropriate education and/or requirements during scientific training), repeats the mistakes of earlier generations (the Santayana principle: "Those who cannot remember the past are condemned to repeat it."). We can find innumerable examples in animal behavior research where scientists: (1) lock too much into theory and not enough into Nature, (2) fail to use all available scientific approaches (do not understand process), (3) seek "truth" as an end product, (4) do not realize that their past experiences can influence perception, (5) fudge results just a little or perhaps emphasize or select "positive" results, (6) Ignore anomalies that arise, (7) seek fame and fail to acknowledge adequately those who went before, (8) elevate some individuals to "hero" status, (9) wallow in anthropomorphism and teleology, and (10) act to exclude (deny) a platform to those with opposing views. Any one or more of these mistakes can lead to controversy and most often do.

But it need not be so. Students of animal behavior can get into the spirit of true science by emphasizing interpersonal relations less and Nature more. In animal behavior studies, the future lies in a greater emphasis on the stimuli responsible for given acts, not on the presumed "function" of those acts—not "Why does a given behavior exist?" but, "What stimulus evokes the behavior?"—just as Schneirla did. Also, we can more often choose the appropriate animal for the question rather than focusing on the animal itself. Finally, we must not strive to obtain certain results in our experiments. If one hopes for a given result, all is lost. Nature has its own rules for us to find, not to dictate.

Our particular community interest in animal behavior studies, of course, is the progress of science and our understanding of Nature, including knowledge of what animals really do. When controversy does erupt—as it surely will, and repeatedly—scientists should not run and hide but rise to the occasion and exploit the spirit and challenge provided.

Acknowledgments

The University of California at Santa Barbara, especially the Department of Biological Sciences in the College of Letters and Science and the College of Creative Studies, provided a haven and opportunity for reflections on the scientific process. Students in my course, "The Nature of Biological Research," also contributed much toward the thoughts contained in this contribution.

I thank Eugene Meyer at Loyola College in Baltimore; William Shurcliff, Emeritus Professor of Physics at Harvard; and Patrick Wells, Emeritus Professor of Biology at Occidental College in Los Angeles, for their valuable suggestions for improvement of the manuscript. Special thanks also go to Gary Greenberg and Ethel Tobach for their role in bringing together the participants in the Sixth T.C. Schneirla Conference.

References

Anderson, J. (1988). Controversies in science: When the experts disagree. *MBL Science, 3,* 18.

Aristotle. (1931). *Historia animalium.* Book 9.40; Vol. 3; Vol. 4. London: Oxford University Press. Original work, 330 B.C.

Aronson, N. (1986). The discovery of resistance: Historical accounts and scientific careers. *Isis, 77,* 630–646.

Asimov, I. (1989, June 9). Cold fusion: Science fiction or reality? *Los Angeles Times.*

Atkinson, J. W. (1985). Models and myths of science: Views of the elephant. *American Zoologist, 25,* 727–736.

Bacon, F. (1952). Novum organum. In R.M. Hutchins (ed.), *Great books of the western world* (pp. 103–195). Chicago: Encyclopedia Britannica, Inc. Original work, 1620.

Bernard, C. (1957). *An introduction to the study of experimental medicine.* New York: Dover. Original work, 1865.

Bernatowicz, A. J. (1958). Teleology in science teaching. *Science, 128,* 1402–1405.

Bernstein, R. J. (1983). *Beyond objectivism and relativism: Science, hermeneutics, and praxis.* Philadelphia: University of Pennsylvania Press.

Bohm, D., & Peat, F. D. (1987). *Science, order, and creativity.* New York: Bantam Books.

Branscomb, L. M. (1985). Integrity in science. *American Scientist, 73,* 421–423.

Broad, W., & Wade, N. (1982). *Betrayers of the truth.* New York: Simon & Schuster.

Bruner, J. S., & Postman, L. (1949). On the perception of incongruity: A paradigm. *Journal of Personality, 18,* 206–223.

Chalmers, A. F. (1978). *What is this thing called science.* Milton Keynes, England: Open University Press. Original work, 1976.

Chamberlin, T. C. (1965). The method of multiple working hypotheses. *Science, 148,* 754–759. Original work, 1890.

Dayton, P. K. (1979). Ecology: A science and a religion. In R. J. Livingston (ed.), *Ecological processes in coastal and marine systems* (pp. 3–18). New York: Plenum Press.

Dewsbury, D. A. (1993). On publishing controversy: Norman R. F. Maier and the genesis of seizures. *American Psychologist, 48,* 869–877.

Dixon, B. (1973). *What is science for?* London: Collins.

Dubos, R. J. (1950). *Louis Pasteur: Free lance of science.* Boston: Little, Brown & Co.

Duclaux, E. (1920). *Pasteur: The history of a mind.* Philadelphia: W. B. Saunders Co. Original work, 1896.

Feyerabend, P. K. (1975). *Against method: Outline of an anarchistic theory of knowledge.* London: New Left Books.

Fleck, L. (1979). *Genesis and development of a scientific fact.* Chicago: University of Chicago Press. Original work, 1935.

Franks, F. (1981). *Polywater.* Cambridge, Mass.: MIT Press.

Gilbert, G. N., & Mulkay, M. (1984). *Opening Pandora's box.* New York: Cambridge University Press.

Glazer, M. P., & Glazer, P. M. (1986). Whistleblowing. *Psychology Today, 20*, 36–43.

Gleick, J. (1992). *Genius: The life and science of Richard Feynman.* New York: Pantheon Books.

Griffith, B. C., & Mullins, N. C. (1972). Coherent social groups in scientific change. *Science, 177*, 959–964.

Griffin, D. R. (1984). *Animal thinking.* Cambridge: Harvard University Press.

Hardin, G. (1993). *Living within limits: Ecology, economics, and population taboos.* New York: Oxford University Press.

Horgan, J. (1990). Stinging criticism. *Scientific American, 26*, 32.

Huizenga, J. R. (1992). *Cold fusion: The scientific fiasco of the century.* Rochester, N.Y.: University of Rochester Press.

Hull, D. L. (1988). *Science as a process: An evolutionary account of the social and conceptual development of science.* Chicago: University of Chicago Press.

Keller, E. F. (1983). *A feeling for the organism: The life and work of Barbara McClintock.* New York: Freeman.

Kuhn, T. S. (1970). *The structure of scientific revolutions.* 2nd ed. enlarged. Foundations of the Unity of Science. Vol. ii, No. 2. Chicago: University of Chicago Press. Original work, 1962.

Kuhn, T. S. (1979). Forward. In Ludwik Fleck (1935/1979), *Genesis and development of a scientific fact.* Chicago: University of Chicago Press.

Latour, B. (1987). *Science in action: How to follow scientists and engineers through society.* Milton Keynes, England: Open University Press.

Lehrman, D. S. (1970). Semantic and conceptual issues in the nature-nurture problem. In L. R. Aronson, E. Tobach, D. S. Lehrman, & J. S. Rosenblatt (eds.), *Development and evolution of behavior: Essays in memory of T. C. Schneirla* (pp. 17–52). San Francisco: W.H. Freeman.

Loehle, C. (1987). Hypothesis testing in ecology: Psychological aspects and the importance of theory maturation. *Quarterly Review of Biology, 62*, 397–409.

Mahoney, M. J. (1976). *Scientist as subject: The psychological imperative.* Cambridge, Mass.: Ballinger (Lippincott).

Maier, N. R. F. (1960). Maier's law. *American Psychologist, 15*, 208–212.

Medawar, P. B. (1979). *Advice to a young scientist.* New York: Harper and Row.

Moore, B. R. (1988). Magnetic fields and orientation in homing pigeons: Experiments of the late W.T. Keeton. *Proceedings of the National Academy of Sciences, 85*, 4907–4909.

Paul, D. B. (1987). The nine lives of discredited data. *The Sciences*, May/June, 26–30.

Pauly, P. J. (1981). *Jacques Loeb and the control of life: An experimental biologist in Germany and America.* 1859–1924. Baltimore: Johns Hopkins University (Ph.D. Thesis, Univ. Microfilm #8106660).

Peters, D. P., & Ceci, S. J. (1982). Peer-review practices of psychological journals: The fate of published articles, submitted again. *Behavioral and Brain Sciences, 5*, 187–255.

Piel, G. (1970). The comparative psychology of T.C. Schneirla. In L.R. Aronson, E. Tobach, D. S. Lehrman, & J. S. Rosenblatt (Eds.), *Development and evolution of behavior: Essays in memory of T. C. Schneirla* (pp. 1–13): San Francisco: W.H. Freeman.

Piel, G. (1984). T. C. Schneirla and the integrity of the behavioral sciences. In G. Greenberg, & E. Tobach (eds.). *Behavioral evolution and integrative levels* (pp. 9–14). Hillsdale, NJ: Erlbaum.

Platt, J. (1964). Strong inference. *Science, 146*, 347–353.

Polanyi, M. (1958). *Personal knowledge: Towards a post- critical philosophy.* Chicago: University of Chicago Press.

Racker, E. (1989). A view of misconduct in science. *Nature, 339*, 91–93.

Rosin, R. (1978). The honey bee "language" controversy. *Journal of Theoretical Biology, 7*, 489–602.

Rosin, R. (1980). Paradoxes of the honey-bee "dance language" hypothesis. *Journal of Theoretical Biology, 84*, 775–800.

Rosin, R. (1988). Do honey bees still have a "dance language"? *American Bee Journal,* *128,* 267–268.

Rosin, R. (1992). A note on the decisive "proof" for use of "dance language" information. *American Bee Journal, 132,* 428.

Rousseau, D. L. (1992). Case studies in pathological science. *American Scientist, 80,* 54–63.

Silvernale, M. N. (1965). *Zoology.* New York: Macmillan Co.

Steinbeck, J. (1962). *The log from the Sea of Cortez.* New York: Viking. Original work, 1941.

Stewart, W., & Feder, N. (1987). The integrity of the scientific literature. *Nature, 325,* 207–214.

Taubes, G. (1993). *Bad science: The short life and weird times of cold fusion.* New York: Random House.

Theocaris, T., & Psimopoulos, M. (1987). Where science has gone wrong. *Nature, 329,* 595–598.

Tobach, E. (1970). Some guidelines to the study of the evolution and development of emotion. In L.R. Aronson, E. Tobach, D. S. Lehrman, & J. S. Rosenblatt (eds.), *Development and evolution of behavior: Essays in memory of T. C. Schneirla* (pp. 238–253). San Francisco: W. H. Freeman.

Tobach, E., & Aronson, L. (1970). T.C. Schneirla: A biographical note. In L.R. Aronson, E. Tobach, D. S. Lehrman, & J. S. Rosenblatt (eds.), *Development and evolution of behavior: Essays in memory of T. C. Schneirla* (pp. xi-xviii). San Francisco: W. H. Freeman.

Wells, P. H., & Wenner, A. M. (1973). Do honey bees have a language? *Nature, 241,* 171–175.

Wenner, A. M. (1972). Incremental color change in an anomuran decapod, *Hippa pacifica* Dana. *Pacific Science, 26,* 346–353.

Wenner, A. M. (1989). Concept-centered vs. organism-centered research. *American Zoologist, 29,* 1177–1197.

Wenner, A. M. (1993). Science as a process: The question of bee "language." *Bios, 64,* 78–83.

Wenner, A. M., & Harris, A. M. (1993). Do California monarchs undergo long-distance directed migration? In S. G. Malcolm & M. P. Zalucki (eds.), *Biology and conservation of the monarch butterfly* (pp. 209–218). Los Angeles: Natural History Museum of Los Angeles County, Science Series contr. No. 38.

Wenner, A. M., Meade, D., & Friesen, L. J. (1991). Recruitment, search behavior, and flight ranges of honey bees. *American Zoologist, 31,* 768–782.

Wenner, A. M., & Wells, P. H. (1987). The honey bee dance language controversy: The search for "truth" vs. the search for useful information. *American Bee Journal, 127,* 130–131.

Wenner, A. M., & Wells, P. H. (1990). *Anatomy of a controversy: The question of a "language" among bees.* New York: Columbia University Press.

Wenner, A. M., Wells, P. H., & Johnson, D. L. (1969). Honey bee recruitment to food sources: Olfaction or language? *Science, 164,* 84–86.

Wilson, E. O. (1972). (Letter exchange). *Scientific American, 227,* 6.

Grades, Levels and Ratchets in Evolution

Niles Eldredge

T. C. Schneirla's concept of "integrative levels" in comparative psychology (e.g., Schneirla, 1949, 1951, 1953), taken together with its major criticism that such an approach is fatally flawed in its reliance on the outmoded "Scala Natura" concept (e.g., Hodos & Campbell, 1969; see Greenberg, in press, for a useful review), has striking parallels within evolutionary theory. My goal in the present discussion is to reveal these parallels and to discuss the major criticisms of the evolutionary concept of *grades* (exact analogues of Schneirla's "levels"). I will conclude that, despite serious conceptual problems, there is indeed a core empirical pattern underlying levels in comparative psychology and grades in evolutionary biology. At least insofar as evolutionary theory is concerned, the baby has been tossed out with the bathwater; something of a *rapprochement* is required, one which I explore at the conclusion of this discussion.

Reading Schneirla, one cannot but conclude that he was very much influenced by his surroundings at the American Museum of Natural History and by developments in comparative biology generally. Intellectual fervor grew in intensity through the 1930s and persisted even after World War II broke out—culminating in the "Modern Synthesis" in evolutionary theory. The term was coined by Julian Huxley (1942)—but the major figures were New Yorkers Theodosius Dobzhansky (e.g., Dobzhansky, 1937) at Columbia University and two colleagues of Schneirla's at the American Museum: Ernst Mayr (e.g., Mayr, 1942) and George Gaylord Simpson (e.g., Simpson, 1944). It was Huxley, too, who coined the term "grade" in a relatively late paper (Huxley, 1958)—and though his analysis was exceptionally clear, it was far more a crystallization of a set of evolutionary precepts prevalent since the 1930s than a completely original formulation. Schneirla (e.g., 1949, 1951, 1953) may not have heard the term "grade" when he conceived his "integrative levels," but there can be no doubt that he was aware of the closely parallel concepts of his contemporary evolutionary biologists.

Evolutionary Grades and Clades

Huxley entitled his 1958 paper "Evolutionary processes and taxonomy with special reference to grades." The paper was, in itself, a useful taxonomy of evolutionary processes. Huxley had three specific evolutionary processes in mind: *anagenesis*, *cladogenesis* and *stasigenesis*. Of these three terms, the first two were immediately snapped up into the everyday language of the evolutionary biological trade. "Anagenesis" simply means the accrual of evolutionary change within lineages; "cladogenesis" means the splitting of lineages. "Stasigenesis," however, never enjoyed wide usage. But *stasis*, of course, has—ever since S. J. Gould and I (Eldredge & Gould, 1972) resurrected stasis as a very general pattern in the evolutionary histories of species. Stasis is simply the tendency of species *not* to display much evolutionary change—to remain, in other words, rather stable throughout their (often protracted) histories. Huxley's stasigenesis would be the "process" underlying such stability—a subject to which I return briefly below.

Cladogenesis produces clades—new lineages; new clades arise through splitting processes epitomized by (allopatric) speciation as developed two decades earlier by Dobzhansky (1937) and Mayr (1942). But, to Huxley, imbued with the very essence of the Darwinian tradition, anagenesis is where the evolutionary action *really* resides. Evolution *is* adaptive change—change under natural selection. Darwin saw selection acting to modify adaptations in two very general contexts. In the first case, as environments change (as was becoming clear they must do given the simple passage of time), natural selection will act to track environmental change, modifying organismic attributes to match the changing conditions—so long, that is, as the requisite heritable variation exists within a species.

But Darwin and many of his intellectual descendants down to the present day also saw linear, directional adaptive change (Huxley's anagenesis) occurring even if the environment itself remains relatively stable. Here the notion of improvement, even of progress, is most evident in the Darwinian canon: selection (once again assuming the existence of the requisite heritable variation) will act to hone existing adaptations, to improve them, given long-term environmental stability—and hence the necessary time for selection to act.

Darwin (1859), as is well known, did not use the term "evolution" in his entire *Origin*—until, that is, the very last word ("evolved"). He spoke, instead, of "descent with modification." Splitting—Huxley's (1958) "cladogenesis"—was a necessary ingredient only because there is (manifestly) more than a single species in the modern biota. But *evolution* is modification—and to Darwin, that modification arises continually throughout the temporal span of species. Indeed, as Mayr (1942) has remarked, Darwin never did address the problem of "speciation" in the modern sense; rather, to Darwin, new species arise simply through the transformation—the linear adaptive change—of their forbears. Species evolve themselves out of existence, transforming from one species to the next in a smoothly gradational manner. Occasionally, the lines might bifurcate (producing the observed multiplicity of lineages), but the real evolutionary action lies in the linear transformation of organismic phenotypic (including behavioral) properties *within* lineages.

Huxley (1958), then, was merely mouthing tradition when he emphasized the importance of anagenesis over cladogenesis (and stasigenesis, for that matter). These themes had been paramount throughout the critical era of the thirties and forties as the Modern Synthesis was emerging. Of the three mathematically inclined geneticists (Ronald Fisher, J. B. S. Haldane and Sewall Wright) who first effected a resolution of the apparent contradictions between the newly developed science of genetics and the older Darwinian view of evolution through natural selection, it was Fisher who came closest to asserting that evolution was simply the change in gene frequencies from generation to generation under the guiding aegis of natural selection. Indeed, Wright (e.g., Wright, 1931, 1932) is most famous for an alternative formulation: genetic drift—whereby genetic differences between populations emerge essentially through chance, and not through differential natural selection.

The synthesis, it is now generally conceded, was considerably more pluralistic in the 1930s than in its later manifestations. Gould

(1980), for example, details the retreat taken by Simpson (1953) from his earlier (Simpson, 1944), more radical stance on macroevolution—embodied in his theory of "Quantum Evolution." In its earlier guise, Simpson had invoked various processes—most notably Wright's genetic drift—to explain how major evolutionary shift can happen abruptly as lineages split. By 1953 Simpson had dropped genetic drift and cladogenesis from his model, referring to quantum evolution as an "extreme and limiting case of phyletic evolution" wholly under the aegis of natural selection.

At the same time, the importance of true speciation, so carefully developed by Mayr and Dobzhansky in the early days of the synthesis, also began to be eclipsed in the post–World War II era. By the time of the Darwinian centennial—1959—evolutionists by and large were unanimous that the original Darwinian perception was indeed the very essence of evolution: generation-by-generation adaptive change within lineages effectively *is* evolution. By 1957 the synthesis (as Gould, 1980, has said) had effectively "hardened" into the very narrow form reflected so clearly in Huxley's (1958) paper. For though he spoke of several sorts of evolutionary process, Huxley clearly set anagenesis ahead of the other two—just as Darwin had done.

But it is for his specification of *grades* that Huxley's paper is largely, and deservedly, remembered. Huxley contrasted grades with clades—and each corresponds to one of his processes: anagenesis yields grades, while cladogenesis produces, of course, clades. Clades are independent lineages. What, then, are grades?

Grades, according to Huxley, are levels of anatomical organization reached as virtual plateaus in the evolutionary process. They are stable adaptive configurations, often linked *seriatim*, deriving from earlier, more primitive levels and often seen as leading to still higher levels.

Grades, produced by anagenesis, of course may occur within single lineages. But it is eloquent testimony to the enthusiasm for a sort of hyperadaptationism that gripped evolutionary biology in the late 1950s and well into the 1960s that the very best examples of grades were thought to be *polyphyletic*: stable plateaus of adaptive configuration were most convincing, most arresting, if arrived at independently in parallel by related, but phylogenetically distinct, lineages.

The distinction is subtle but important: vertebrate powered flight was never, to my knowledge, considered an adaptive, evolutionary grade. Birds, bats and pterosaurs, while all amniote vertebrates, have long been considered simply too distantly related to constitute a coherent grade. Vertebrate powered flight is (or so it was proclaimed) an example of convergent evolution. Parallel evolution, in contrast, was held to occur within more closely related taxa—where similar mutations and selection pressures might be imagined to produce closely similar evolutionary results, albeit independently.

Thus grades could be monophyletic but were preferably "polyphyletic." And lines, however vague, were drawn between putative examples of true parallelism and the coarser phenomenon of convergence. The very vagueness of these terms, as we shall shortly see, led to one of the major, most devastating critiques of the entire notion of grades—effectively consigning the concept in the past 20 years to the scrap heap of evolutionary theory.

But, while in vogue, the notion of grades provided the focus for a great deal of evolutionary theory and research. No less a figure than George Gaylord Simpson enthusiastically embraced grades in his analysis (Simpson, 1959a) of the origin of Class Mammalia. Mammals constitute a grade of organization—typified by the possession of hair, mammary glands and 3 middle ear bones—reached not once but several times independently in the course of therapsid ("reptilian") Mesozoic history—or so Simpson concluded. In a separate paper (Simpson, 1959b), Simpson broached the concept of "key innovations"—tracing adaptive radiations to evolutionary novelties that set the stage for adaptive diversification of new taxa. The two concepts—grades and key innovations—work together quite well, in that theorists were fond of postulating "radiations" occurring within each level in a series of gradal evolutionary steps. Suffice it to say, the notion of key innovations has survived as a useful evolutionary theoretical rubric—primarily because its effects can be imagined wholly within the context of monophyly. The popularity of grades hinged to such a great degree on the concept of polyphyly that grades have disappeared as polyphyly has come to be viewed more as a sign of deficient analysis than as a true evolutionary mode.

Simpson's views on the polyphyletic origins of mammals were quickly laid to rest. But there were many other examples of grades prevalent in the literature in the days surrounding the centennial of the *Origin*. Most notorious, perhaps, was Carleton Coon's (e.g., Coon, 1962) resurrection of the polyphyletic origin of our own species, *Homo sapiens*. Somehow, a single, reproductively coherent species was supposed to have arisen at least three separate times, as the advanced properties that mark our species and set it off from more primitive fossil forbears were held to have been developed independently in Africa, Europe and Asia. And though, curiously, Coon's views are still very much alive and well (e.g., Wolpoff, 1986), by and large the theoretical precepts and analytical procedures of modern evolutionary biology are finally beginning to reach even paleoanthropology (Tattersall, 1993).

There are many other examples. A major symposium (Schaeffer & Hecht, 1965) was very likely the last word—in more than one sense—on the subject. Bobb Schaeffer and Max K. Hecht are also American Museum paleontologists—further testimony to the importance attached to gradal notions at that institution—notions that were to last there to the mid-1960s. Significantly, though, it was precisely these two paleontologists who (together with myself as a very junior author) wrote the first *positive* exegesis of "cladistics" (originally known as phylogenetic systematics—Hennig, 1966) from a paleontological viewpoint (Schaeffer, Hecht & Eldredge, 1972). And it was cladistics, of course, that provided one of the two mortal blows to the popularity of grades as a focus of concern in evolutionary biology generally.

Grades and Phylogenetic Systematics: The First Critique

"Cladistics" was a somewhat derogatory term coined by Ernst Mayr (1974) in an early attempt to stave off the growing influence of what was at the time a novel approach to phylogenetic analysis and systematics. Mayr was complaining that, in strictly focusing on lineages ("clades"), the new technique ignored all the other components of evolution—components already well integrated in the "new systematics" (Huxley's term, but more Mayr's doing).

And Mayr was certainly right in his basic characterization of cladistics: cladistics is the formalized search for lineages, for taxa composed of a single ancestral and all later descendant species. After all, evolution produces lineages if it produces anything. Evolutionary history is genealogical history. And, in an elegant paper, Platnick and Cameron (1977) showed the formal identity in principles of genealogical research shared by cladistics, and the earlier-developed fields of historical linguistics and stemmatics (i.e., textual criticism, the comparative study of alternative versions of manuscripts, including their interrelationships—where apparently these principles were first clearly codified).

Nor was the point lost on early adherents that, in a very real sense, cladistics merely codifies and makes explicit what "good" systematists had been doing all along: looking for "natural" groups—ever since Darwin, understood as genealogically pure taxa. For example, it was a realization from pre-Darwinian times that porpoises, sharks and the extinct ichthyosaurs, however closely their fusiform bodies might resemble each other, do not constitute a "natural" group. Porpoises are mammals, sharks chondrichthyans and ichthyosaurs reptiles.

The goal of cladistics is to define and recognize only monophyletic groups, ferreting out and correcting less obvious examples of non-phylogenetically coherent taxa. Two principal sorts of nonmonophyletic taxa were soon distinguished: paraphyletic taxa (in which not all known descendants of an ancestral species are included, often implying that some included taxa are more closely related to other taxa *not* included within the group); and polyphyletic taxa (in which taxa descended from more than one single ancestral species are included). The following example, in which paraphyletic, monophyletic and polyphyletic taxa are clearly distinguished, also sheds light on the nature—and fate—of grades.

Whittaker's (1959) precladistics "five-Kingdom" classificatory scheme of the major divisions of life contains examples of all three kinds of taxa. The five kingdoms are Bacteria, Protista, Plantae, Fungi and Animalia. But there is additional phylogenetic structure in the scheme: according to Margulis (a major champion of Whittaker's classification—see Margulis & Schwartz, 1988), the Bacteria are the sole members of a still higher taxon, the Prokaryota. Correspondingly, the Eukaryota contain the remaining four Kingdoms.

Monophyly is established by positive evidence: sets of attributes unique to a taxon, uniting its members. To show that such a group is polyphyletic, one must demonstrate that such resemblances are not homologous—by demonstrating that some constituent groups actually share derived features with other groups. Paraphyletic groups, on the other hand, typically lack any definitively, uniquely defining characteristics at all.

Prokaryotes are a textbook example of a paraphyletic group. Prokaryotes are defined on the basis of the *absence* in them of the defining properties of Eukaryota: to name but one, eukaryotes are united by the presence of a double-walled membrane surrounding a discrete nucleus housing chromosomes. Prokaryotes lack such structures: prokaryotes are all those organisms that are not eukaryotes. No surprise, then, that Woese (1981) has shown that prokaryotes are actually quite diverse; some bacterial groups appear to be more closely related to eukaryotes than other bacteria. Bacteria are hardly a unified grade—and to a good phylogenetic systematist, Prokaryota hardly constitute a natural evolutionary group.

But Eukaryota, with its extensive list of shared characters, are generally conceded to constitute a true monophyletic group. Bacteria (Prokaryota) constitute a paraphyletic grade; eukaryotes, too, are a grade, a level of cellular anatomical organization—but they are also a *bona fide* natural, evolutionary group. The two-level succession of the two grades loses focus in the eyes of many with the nonmonophyletic status of Prokaryota.

The Eukaryota yield another two examples of the same general phenomenon. The protistans are conventionally defined as single-celled eukaryotes—all those eukaryotes that are neither multicellular fungi nor animals nor plants. As simply primitive eukaryotes, protists can confidently be expected to be variously allied, with some sharing phylogenetic connections with plants, some with animals, some with fungi—while some, of course, are not particularly closely related to any of the multicellular taxa.

Moreover, multicellularity is manifestly a grade-group: no one seriously defends recognition of an evolutionary group consisting of all multicellular eukaryotes—the equivalent of recognizing a shark+porpoise+ichthyosaur group. There has just been too much evidence for too long that, for example, certain protists with photosynthetic capabilities are the nearest relatives to plants, others to fungi and still others to animals. Thus no monophyletic taxon consisting strictly of multicellular eukaryotes can rationally be recognized.

In the eyes of Julian Huxley and many of his contemporaries, the very strength of the concept of grades was their plurality of composition: sure, grades *might* be monophyletic—but they just as likely might be polyphyletic (paraphyly was essentially unrecognized). Indeed, it was generally considered preferable if grades *were* polyphyletic—the more to emphasize the intense power of the anagenetic process of adaptation through natural selection. The strong revulsion against trends that accompanied the rise of cladistics in the late 1960s stemmed from the view that evolution is fundamentally a genealogical process and produces monophyletic skeins of taxa. Everything else is imaginary. Precisely because nonmonophyly was openly embraced as a *good thing* insofar as grades were concerned, the gradal concept was deeply eclipsed when systematics underwent its much-needed revamping during the cladistics revolution.

Grades and Evolutionary Theory: The Second Critique

The ineluctable power of natural selection: Such was the revived legacy of Darwin, who changed the world from thinking that evolution is impossible to seeing evolution as inevitable given the mere passage of time. Time passes, environments change and natural selection willy-nilly, given only the requisite variation, will track that environmental change. The concept of grades comes directly out of this intellectual legacy.

But it simply is not so: True, environments inevitably do change. But the time-honored imagined consequence, it turns out, was ecologically, even commonsensically, naive. Environmental change—as when glaciers invade lower latitudes periodically, bringing the tundra and boreal forest down in front of them—does indeed engender tracking. But it is not the sort of tracking that evolutionary biologists traditionally have had in mind. Rather than selection directionally modifying the traits of organisms to meet changing conditions, organisms (hence populations and therefore entire species) change their distributions. Species ranges are constantly changing in response to climatic and other forms of environmental change. This is habitat tracking—a very different sort of affair than Darwin imagined.

As long as a species can continue to find and occupy recognizable habitats—habitats to which their economic and reproductive adaptations are functionally suited—that species will persist. The alternative to habitat tracking, it turns out, is not so much gradual evolutionary modification to meet the new set of environmental parameters, but rather extinction. Most species that have ever existed are now extinct—victims of habitat loss.

Moreover, there is now abundant empirical evidence that, so long as a species can find a suitable habitat, it will tend to persist more or less unchanged. Most species vary geographically. It has been a major realization of the past 2 decades that species characteristically display little more total variation throughout their entire temporal histories (which are often in the millions of years) than they do at any one time spread out over a heterogeneous landscape. This great species stability (*stasis*), the underlying empirical percept of punctuated equilibria (Eldredge, 1971; Eldredge & Gould, 1972) was known to some of Darwin's contemporaries (especially paleontologists). Darwin himself mentions stasis in passing in his sixth edition of the *Origin*—where he is busily engaged in countering the various objections that had been raised to his initial discussion of evolution.

Suffice it to say that, increasingly, adaptive change is seen to be concentrated in relatively brief periods, most commonly in association with the actual cladogenetic process of speciation itself. Progressive adaptive change is far from the hallmark of the histories of species. We have had to abandon the notion that gradual, progressively adaptive evolutionary change is the primary "signal" in the evolutionary history of life.

I hasten to add that adaptation is alive and well as a process/ state in evolutionary theory. But much of the uncritical enthusiasm for the all-mighty workings of natural selection and the ubiquity and constancy of adaptive transformation has recently been heavily criticized for a host of methodological and theoretical reasons (see Gould & Lewontin, 1979, for an apt and well-known review). Such critiques are in many ways understandable reactions to the unbridled enthusiasm for an excessively narrow construal of the nature of the evolutionary process—an enthusiasm that peaked in a great crescendo in the late 1950s.

But there is one further theoretical point, one based on firm empirical evidence and solid analysis—and one that has a direct bearing on the issue of grades. I refer to the issue of "historical contingency" in evolution—an argument developed especially cogently by S. J. Gould in recent years (e.g., Gould, 1989).

In the American version of the movie *Godzilla vs. King Kong*, Tokyo (and, ultimately, the entire world) is threatened by a gigantic, fire-snorting monster: Godzilla, a marvelous blend of *Tyrannosaurus rex* and medieval dragons. Powerless to stop the onslaught, human fate hinges on the success of King Kong. Mighty in his own right, though, the overgrown ape is no physical match for Godzilla. Things, at first, don't go at all well. But, in due course, (inevitably?) King Kong prevails. It all boils down to brains over brawn: Kong manages to outwit the dim brute in the end.

This grade-B movie plot is actually a pretty fair trot on traditional thinking about the fate of dinosaurs, the rise of mammals—and, more generally, the relation between extinction and evolution. Dinosaurs were great in their time, but finally they had to give way to us superior mammals. Darwin and most of his descendants have—until recently—persisted in seeing extinction as a side effect of evolution: as newer, superior evolutionary products emerge, the old simply give way, unable to keep up with the newer, better competition.

But it simply does not work that way. Mammals (as had been known in the mid-nineteenth century) appeared alongside the dinosaurs in the Upper Triassic. For reasons unknown, it was the dinosaurs, rather than the mammals, that radiated to fill the tetrapod vertebrate components of terrestrial ecosystems. They (as birds) and collateral kin (pterosaurs) developed powered flight; and other reptilian lineages (mosasaurs, plesiosaurs, ichthyosaurs) invaded the waterways (along with crocodilians and turtles). Several times—most notably, in the great mass extinction episode that ended the Triassic Period—dinosaur numbers were severely depleted. But after each such episode during the 170-million–odd years that dinosaurs and collateral reptilian kin were the dominant tetrapods in the world's ecosystems, it was these groups, *not* the mammals, that would once again diversify as ecosystems were reformed.

It was only after the extinction event at the end of the Cretaceous Period some 65 million years ago—an event that finally saw the total eradication of all existing dinosaur species—that the mammals literally inherited the earth. With no dinosaurs to radiate back into terrestrial ecosystems, the mammals finally had their chance. After a lag of several millions of years, for the first time in 170 million years of living, mammals evolved into an array of different body sizes and trophic preferences: large herbivores, various carnivores and so forth.

That is Gould's contingency: no one can ponder the literal record of dinosaurs and mammals and seriously maintain that mammals were destined to rise up and smite the dinosaurs because of some form of latent superiority that would, like King Kong, win out in the end. One cannot help but think, instead, that had that end-Cretaceous event not completely devastated the dinosaurian fauna, they would still be here—and we wouldn't be contemplating issues in evolutionary biology and comparative psychology. Indeed, we wouldn't be here at all.

What a change from the traditional credo of the inevitability of progressive adaptive change! We have adjusted our thinking radically on the nature and context of adaptive change in evolution. Grades, of course, were the very embodiment, a literal incarnation, of such thinking. And now we are rapidly growing accustomed to thinking in very different terms. In a scant 30 years, grades have all but vanished from evolutionary discourse. And that, in a way, is a shame: for there are issues left unexamined if we turn our backs completely on the concept of levels and grades.

Toward Rapprochement: Signal and Noise in Evolutionary Pattern

All of us in the natural sciences are concerned with developing explanations of natural phenomena. To do so, of course, demands scrupulous attention first, and foremost, to the nature of those entities that occupy the natural universe. We then seek to understand what these entities *do*. Repeated patterns of events are especially crucial to those engaged in the study of historical processes. Allopatric speciation, punctuated equilibria, current theory on the relation between extinction and evolution, and Simpson's quantum evolution are all examples of process theories firmly rooted on repeated, generically similar events: historical patterns.

I have argued, in effect, that much of Darwinian theory, when applied to long-term entities, events and processes, has been overly extrapolationist. The kinds of patterns elicited in the laboratory and by selective breeding have been uncritically extrapolated across the board, unfettered in any systematic way by the facts of the matter—the sorts of patterns with which we should actually begin our analyses of nature of the evolutionary process in real, historical time.

But that is theory. Grades—levels of organization, be they anatomical or behavioral (as in, of course, the case of T. C. Schneirla's conceptualization)—fall into the realm of actual pattern: there is signal in these grades, something that definitely needs to be explained.

But what manner of "pattern" are these grades? One major reason for the eclipse of gradal thinking, as we have seen, came from presenting grades literally as real entities—specifically as "taxa." For good and profoundly simple reasons, we now understand taxa as real entities—historical skeins of interrelated species. But only monophyletic taxa are "real" in that sense. True grades *could* be monophyletic (in Huxley's taxonomy), but that isn't what makes them interesting. In admitting polyphyletic "taxa" into the gradal taxonomy, proponents of grades did themselves no favor.

Nor, of course, are grades "real" if they are imagined to be way stations (or plateaus) along a spectrum of primitive-derived phyletic transformation. Such constructs face the traditional criticism (even from proponents) of lack of discrete, formal boundaries separating such levels.

If grades are not real, spatiotemporally bounded entities unless they happen also to be monophyletic taxa—what are they anyway? Nothing, as many evolutionists seem to have concluded in recent years? What, after all, were serious students—including many empiricists, like Schneirla—describing when they thought in terms of levels or grades?

One element of traditional discourse that if anything has been revived in recent years in evolutionary biology—particularly in the canon of cladistics—is the paired, albeit relative, notions of "primitive" and "derived." In the 1960s, when I was a student, such terms were avoided as somehow value laden—almost as if they were not "politically correct" in some way. We now use the terms as a matter of course: a character is "primitive" if it represents a prior state to some later condition. It is no longer a pejorative, as we have largely abandoned equating "primitive" with "inferior," and "derived" with "superior." (Contrasting "primitive" with "derived" instead of with the more traditional "advanced" has helped to establish that break).

And evolution, by its very nature, is awash with primitive and derived states. That's what evolution—descent with modification—*is*. Previous treatments of grades emphasized anagenesis to the near exclusion of cladogenesis. But it is precisely because cladogenesis is so important in evolution, and also because later groups do not systematically expunge earlier groups (because of some vaunted "superiority"), that we can see the actual signal underlying the gradal (and levels) rhetoric of several decades ago. Cladogenesis and persistence of the primitive yield a spectrum of primitive-derived states in existing taxa: the history of life is all around us today, and gone are any thoughts that the fossil record is the *only* way that we can sketch the outlines of phylogenetic history.

Gone are the value judgments of "superior." What is left is that dizzying multivariate morphocline, that spectrum of primitive and derived states in the biota. Gone is a sense of progress, but with us remains the undeniable spectrum of simple to relatively more complex (with the vector occasionally reversed, as in the simplified bodies of many parasitic organisms). There are undeniably levels of complexity in anatomical organization: multicellularity, which has appeared at least thrice in evolutionary history, by definition is a level of greater complexity than unicellularity (a sponge has several cell types; an amoeba only one). Coelenterates have true tissues (sponges lack them) yet lack the complex internal organs of "higher" Metazoa.

And so forth. Most spectra of complexity involve proliferation of subcomponents in a division-of-labor sense. Insect societies display such spectra (cf. Wilson, 1985). I have no doubt that behavioral attributes, like phenotypic properties generally, usefully and objectively also display such spectra. It is an arresting feature of all such systems with which I am familiar that the primitive, less complex systems habitually coexist with the more complex.

These primitive-derived spectra are the wheat that, separated from the chaff of "progress," "superiority" and so forth, leave patterns of relative complexity which can usefully be compared with one another. How is it, exactly, that complex forms evolve, yet do not displace earlier taxa?

Evolution is much like a ratchet: new and often more complex systems occasionally appear, but they are tacked on to the array of preexisting systems. That is the true significance of cladogenesis. Indeed, without cladogenesis, there would be no anagenesis, or so the ubiquity of within-species stasis would seem to imply. Speciation, the very foundation of cladogenesis, triggers adaptive change in ways not yet fully understood (Eldredge, 1995). At the very least, cladogenesis isolates segments of genetic information, injecting whatever evolutionary novelties that may be present in fledgling species into the phylogenetic mainstream. Without the partitioning effect of speciation, novelty within species tends to be lost—suppressed by the flux within and between populations, as Wright (e.g., 1931) showed so long ago.

Anagenesis lives, but it is ensconced within cladogenesis. Levels and grades are useful concepts, but only if we do not insist that they conform to real taxa, seeing them instead as plateaus within spectra of primitive-derived states—especially, if not exclusively, in the context of relative complexity.

References

Coon, C. S. (1962). *The origin of races.* New York: Knopf.

Darwin, C. (1859). *On the origin of species.* London: John Murray.

Dobzhansky, T. (1937). *Genetics and the origin of species.* Reprinted, 1982. New York: Columbia University Press.

Eldredge, N. (1971). The allopatric model and phylogeny in Paleozoic invertebrates. *Evolution, 25*, 156–167.

Eldredge, N. (1995). *Reinventing Darwin. The great debate at the high table of evolutionary theory.* New York: John Wiley & Sons.

Eldredge, N., & Gould, S. J. (1972). Punctuated equilibria: an alternative to phyletic gradualism. In T. J. M. Schopf (ed.), *Models in Paleobiology* 51 (pp. 82–115). San Francisco: Freeman, Cooper.

Gould, S. J. (1980). G. G. Simpson, paleontology, and the modern synthesis. In E. Mayr & W. B. Provine (eds.), *The evolutionary synthesis: Perspectives on the unification of biology* (pp. 153–172). Cambridge, Mass.: Harvard University Press.

Gould, S. J. (1989). *Wonderful life. The Burgess shale and the nature of history.* New York: W.W. Norton.

Gould, S. J., & Lewontin, R. C. (1979). The spandrels of San Marco and the Panglossian paradigm: A critique of the adaptationist programme. *Proceedings of the Royal Society of London, B 205*, 581–98.

Greenberg, G. (1995). Anagenetic theory in comparative psychology. *International Journal of Comparative Psychology, 8*, 31–41.

Hennig, W. (1966). *Phylogenetic Systematics.* Urbana: University of Illinois Press.

Hodos, W., & Campbell, C. B. G. (1969). *Scala naturae*: Why there is no theory in comparative psychology. *Psychological Review, 76*, 337–350.

Huxley, J. S. (1942). *Evolution: The modern synthesis.* New York: Harper.

Huxley, J. S. (1958). Evolutionary processes and taxonomy with special reference to grades. *Uppsala University Arsskrift*, 21–38.

Margulis, L., & Schwartz, K. V. (1988). *Five kingdoms*. 2nd ed.. New York: W.H. Freeman.

Mayr, E. (1942). *Systematics and the origin of species*. Reprinted, 1982. New York: Columbia University Press.

Mayr, E. (1974). Cladistic analysis or cladistic classification? *Zeitschrift für Zoologische Systematik und Evolutionforschung, 12*, 94–128.

Platnick, N. I., & Cameron, D. H. (1977). Cladistic methods in textual, linguistic, and phylogenetic analysis. *Systematic Zoology, 26*, 380–385.

Schaeffer, B., & Hecht, M. K. (eds.) (1965). *Symposium: The origin of higher levels of organization. Systematic Zoology*, 245–342.

Schaeffer, B., Hecht, M. K., & Eldredge, N. (1972). Phylogeny and paleontology. *Evolutionary Biology, 6*, 31–46.

Schneirla, T. C. (1949). Levels in the psychological capacities of animals. In R. W. Sellars, V. J. McGill & M. Farber (eds.), *Philosophy for the future: The quest of modern materialism* (pp. 243–286). New York: MacMillan.

Schneirla, T. C. (1951). The "levels" concept in the study of social organization in animals. In M. Sherif & J. N. Rohrer (eds.), *Social Psychology at the Crossroads* (pp. 83–120). New York: Harper.

Schneirla, T. C. (1953). The concept of levels in the study of social phenomena. In M. Sherif & C. Sherif (eds.), *Groups in harmony and tension* (pp. 54–75). New York: Harper.

Simpson, G. G. (1944). *Tempo and mode in evolution*. New York: Columbia University Press.

Simpson, G. G. (1953). *The major features of evolution*. New York: Columbia University Press.

Simpson, G. G. (1959a). Mesozoic mammals and the polyphyletic origin of mammals. *Evolution, 13*, 405–414.

Simpson, G. G. (1959b). The nature and origin of supraspecific taxa. *Cold Spring Harbor Symposium Quantitative Biology, 24*, 255–271.

Tattersall, I. (1993). *The human odyssey. Four million years of human evolution*. New York: Prentice Hall General Reference.

Whittaker, R. H. (1959). On the broad classification of organisms. *Quarterly Review of Biology, 34*, 210–226.

Wilson, E. O. (1985). The sociogenesis of insect colonies. *Science, 228*, 1489–1495.

Woese, C. R. (1981). Archaebacteria. *Scientific American, 224*, 92–122.

Wolpoff, M. (1986). Describing anatomically modern *Homo sapiens*. A distinction without a definable difference. *Anthropos, 23*, 41–53.

Wright, S. (1931). Evolution in Mendelian populations. *Genetics, 16*, 97–159.

Wright, S. (1932). The roles of mutation, inbreeding, crossbreeding, and selection in evolution. *Proceedings of the Sixth International Congress of Genetics, 1*, 356–366.

Where Have I Heard It All Before

Some Neglected Issues of Invertebrate Learning

Charles I. Abramson

A major force guiding much of the contemporary work on invertebrate learning is the use of what has become known as the simple system approach. In this approach invertebrates are subjected to various nonassociative and associative conditioning paradigms in the hope that the cellular basis of the behavioral change may be revealed. Nonassociative learning is a form of behavior modification in which the presentation of a stimulus leads to an altered strength or probability of a response according to the strength and temporal spacing of that stimulus. This type of behavioral change is studied by using procedures that generate habituation and sensitization. In contrast to nonassociative learning, associative learning is generally considered more complex because the range of what can be learned is vastly expanded. For example, an association can be formed between two or more stimuli, two or more responses, or between a stimulus and a response. This type of behavioral change is studied by using procedures that generate classical, instrumental and/or operant conditioning.

The importance of the simple system approach lies in the possibility of identifying neuronal circuitry and cellular changes associated with learned behaviors. Nonassociative and associative paradigms that have been explored in these simple systems include habituation, sensitization and classical conditioning of gill withdrawal in *Aplysia;* classical conditioning and conditioned suppression of phototaxis in *Hermissenda*; and classical conditioning of proboscis extension in insects. There have also been several examples of application of operant methods, notably conditioning of head waving in *Aplysia* and conditioning of leg position in the locust and several species of crab. Recent reviews of this rather extensive literature are available (Abramson, 1994; Squire, 1992).

The simple systems approach is the dominant focus of most of the contemporary research conducted with invertebrates. Research with invertebrates should not be seen as only a simple system vehicle. As this volume demonstrates there is much that the study of invertebrate behavior can reveal, for instance, about the learning process, foraging, socialization and exploratory behavior.

The purpose of this chapter is to identify five behavioral issues and needs that must be addressed if the "simple system" approach to invertebrate learning is to remain vital. Some of these issues apply to the study of vertebrate learning as well. These issues and needs are:

1.   The extent of phyletic differences in vertebrate and invertebrate learning phenomena needs to be determined;

2.   Inconsistencies in the definitions of learning phenomena are an unresolved issue;

3.   Taxonomies of invertebrate learning paradigms should be developed;

4.   The reporting of individual level data, as contrasted with species or other population classifications should be encouraged;

5.   Cognitive explanations of invertebrate learning phenomena presents a problematic issue.

The needs, of course, lead to issues and vice versa. However, there should be some uniform approach to all five of these. Consider, for example, point 4. Because there is not enough attention paid to variability within species, statements are made about species characteristics that are not valid or reliable. This could be dealt with if individual data were reported. Consider, also, point 5. By employing cognitive explanations of invertebrate learning phenomena, significant species differences are overlooked and conclusions about the equivalence of the behavior of invertebrates and vertebrates are reached. In the extreme, for example, ants and humans are both seen as information-processing systems. The need is to understand the similarities and differences between ants and humans not in terms of language borrowed from human information processing but in terms of sensory integration, perception of stimuli, sensitivity to environmental contingencies and evolutionary history.

These needs and issues have been recognized, in one way or another. Bitterman (1975) discussed the need to consider the role of evolution in learning research. Gormezano (1984) reminded us that what are thought to be classical conditioning procedures may not be classical at all, and Schreurs (1989) has taken this a step further and suggests that classical conditioning in *Aplysia* may not be Pavlovian. Staddon and Bueno (1991) counseled that an understanding of behavior must precede the understanding of brain-behavior relationships. Hirsch and Holliday (1988) warned about the need to report individual data. Amsel (1989) rejected the trend to characterize invertebrate behavior as cognitive, and Tavolga (1969) has argued that ". . . any attempt to interpret the behavior of an insect with concepts and methods based on human psychology would be as ridiculous as attempting a Freudian psychoanalysis of a cockroach" (p. 21).

Anatomists, biochemists, biologists, ethologists, entomologists, geneticists, naturalists, neuroethologists, psychologists, physiologists and zoologists have given us an understanding of the neuronal circuitry and cellular changes associated with behavioral plasticity. However, invertebrates as an example of simple systems are seldom studied for what they can tell us about the learning process as complete organisms; only their nervous systems are used for what they can tell us about the learning process. The invertebrate learning literature should and must be improved by reducing the large gaps in our practical knowledge of behavior, especially learned behavior.

The importance of gathering parametric data is clear. Parametric data have contributed directly to the construction and testing of theories to account for vertebrate behavior (Hull, 1943), established training parameters for a number of vertebrate learning techniques (Sidowski, 1966) and facilitated qualitative and quantitative comparisons among vertebrate species (Macphail, 1987). This rigorous application of parametric analysis is missing from the contemporary analysis of invertebrate learning.

Conceptual behavioral issues discussed in the vertebrate literature include problems posed by (1) the concept of the operant (Catania, 1973; Lee, 1988; Schick, 1971; Staddon & Zhang, 1989); (2) the definition of a behavioral response (Bolles, 1983; Schoenfeld,

1976); and (3) the creation of "behavioral units" to facilitate the analysis of behavior (Thompson & Zeiler, 1986). Such discussions of behavioral issues are less frequent in the contemporary invertebrate learning literature. In general there is a need for greater critical thinking in *all* our work, particularly about the pervasive problems that need to be resolved by theoretical discourse and empirical study.

Issues and Needs in the Analysis of Invertebrate Learning

The Extent of Phyletic Differences in Vertebrate and Invertebrate Learning Phenomena Needs to Be Determined

Schneirla (1949) and Beach (1950) criticized the lack of an evolutionary perspective in animal learning, that would consider differences as well as similarities in the behavior of different species. Classic textbooks in comparative psychology suggest qualitative and quantitative differences in learning ability between invertebrates and vertebrates (e.g., Maier & Schneirla, 1935/1964; Thorpe, 1956; Warden, Jenkins & Warner, 1940; Washburn, 1908). Schneirla, for instance, in a series of classic experiments on learning in ants reported "striking differences" in the learning ability of insect and rat. This early literature is all but forgotten in contemporary reviews of invertebrate learning. Contemporary examples that have attempted to stress differences are the work of Bitterman and his associates (1988), and the work of Lejeune and her associates (Lejeune & Wearden, 1991), showing phyletic differences in learning.

Now, the invertebrate simple system literature appears to *minimize* differences in learning between invertebrates and vertebrates. For example, Sahley (1984) states ". . . it is clear that striking commonalities exist between learning in invertebrates and learning in vertebrates" (p. 190). I have made similar statements in my own work (Abramson, 1986).

Such statements are premature, if not wrong, in their emphasis and require that behavioral scientists gather supporting data on all aspects of invertebrate behavior and as it differs from that of vertebrates as well.

Basic questions present themselves. For example, do many of the invertebrate simple system preparations actually exhibit classical conditioning? Are operant conditioning and avoidance conditioning as widespread among all phyla as believed? The assumption that there is a neural or molecular commonality underlying learning is an assumption that still needs support. Learning has been explored in so few invertebrate species that clearly there is a need to ask about the extent to which the results can be generalized to different behaviors and across phyla.

A fundamental question is: Where does such a comparative analysis begin and how should the analysis be accomplished. An examination of the various standardized preparations whose development was stimulated by the simple system approach shows that many of these are methodologically sophisticated. Stimuli and experimental contingencies are presented automatically and have been modified according to the characteristics of the animals. These include proboscis extension in bees (Bitterman, Menzel, Fietz & Schäfer, 1983), houseflies (Fukushi, 1979) and blowflies (Akahane & Amakawa, 1983); leg lift conditioning in cockroaches (Willner, 1978), locusts (Horridge, 1962), fruit flies (Booker & Quinn, 1981), crickets (Zaldivar & Jaffé, 1987) and crabs (Dunn & Barnes, 1981); lever pressing in cockroaches (Rubadeau & Conrad, 1963), bees (Pessotti, 1972), crayfish (Olson & Strandberg, 1979), crabs (Abramson & Feinman, 1990b), *Aplysia* (Downey & Jahan-Parwar, 1972) and *Helix* (Balaban & Chase, 1989).

There are a number of comparative strategies that can be used in combination to describe the similarities and differences among invertebrate preparations (Brookshire, 1970). One method is to compare the speed or accuracy of learning of a similar response. For example, the crab, fly, locust and cockroach can all be compared on a leg lift task. A second method is to vary systematically the parameters of training variables such as the conditioned-unconditioned stimulus (CS-US) interval, probability of reward and stimulus intensity. Both methods will provide quantitative descriptive data, but qualitative data can also be obtained by searching for the presence of a particular type of learning. For instance, a comparison of solitary and social insects might reveal that the ability to perform in a signaled avoidance procedure is more readily expressed in the social insects.

It is important to keep in mind in conducting any comparison that the differences in abilities among organisms may be the result of some procedural or organismic variable such as motivation, sex, age, sensory and motor apparatus and/or training parameter. These and other variables must be systematically explored before it can be said with any confidence that a quantitative or qualitative difference among any species has been uncovered.

Inconsistencies in the Definitions of Learning Phenomena Are an Unresolved Issue

Within the area of invertebrate learning is the lack of consensus among researchers as to the definitions of many invertebrate learning phenomena. For example, the popular olfactory conditioning situation developed for fruit flies (Quinn, Harris & Benzer, 1974) has been considered to represent classical conditioning (McGuire, 1984), signaled avoidance (Dudai, 1977) and instrumental conditioning (Tully & Quinn, 1985). This lack of consistency has been such a problem in the past that it was addressed in two workshops (Bullock, 1966; Teyler, Baum & Patterson, 1975). To illustrate this lack, examples will be selected from studies of classical conditioning, operant conditioning and signaled avoidance conditioning.

Classical Conditioning. Gormezano and Kehoe (1975) and Gormezano, Kehoe and Marshall (1983) describe four variations of classical conditioning based on the nature of the conditioned response (CR): (1) Conditioned Stimulus–Conditioned Response (CS-CR), (2) Conditioned Stimulus–Instrumental Response (CS-IR), (3) Instrumental Approach Behavior and (4) Autoshaping.

1. The Conditioned Stimulus–Conditioned Response paradigm is considered by Gormezano and Kehoe to represent the "pure" case of classical conditioning. Here the CS does not elicit the unconditioned response (UR) prior to training, and the CR emerges from the same effector system as the UR.
2. The Conditioned Stimulus–Instrumental Response paradigm contains those experimental designs commonly known as "transfer of control" or "classical-instrumental transfer" in which classical conditioning is assessed not directly but by its influence on instrumental or operant responding. Perhaps the most well-known example of

this design is the conditioned suppression procedure. In one version of the conditioned suppression procedure, an animal is placed in an apparatus where it receives several pairings of tone (CS) and shock (US). The effectiveness of the tone-shock pairing is not assessed by looking for a CR after each pairing but rather during a second experiment in which the animal has previously been taught to press a lever for food. During this lever press experiment, the CS is now presented, and the extent to which it disrupts the pattern of lever pressing is assessed.

3. The Instrumental Approach design is a type of conditioning in which some approach behavior to a CS is necessary to receive the US. This procedure is illustrated by general activity to stimuli preceding food (i.e., conditioning of general activity) and instrumental runway and maze situations in which movement toward the food source is necessary.

4. Autoshaping is related to the CS-CR paradigm with the interesting property that the CR is not from an effector system that is related to the US. All of these four categories differ in many ways: how the CR is measured; the accuracy with which the CS and US are presented in a response independent fashion; the nature of the target response; the amount of control the experimenter has over the training variables; and the degree to which the animal is restrained in the conditioning situation.

In their discussions of classical conditioning, Gormezano and Kehoe offer a classification, or taxonomy, that is useful not only as a definition of classical conditioning but also as a point of comparison with what is considered to be classical conditioning in the invertebrate literature. It is important to note that Gormezano and his colleagues consider the CS-CR paradigm the only unambiguous case of classical conditioning. As Gormezano uses the term, and according to their taxonomy, most invertebrate researchers may not be studying classical conditioning. It is interesting to note that almost 40 years ago, classical conditioning of planarians (e.g., Thompson & McConnell, 1955) was largely discredited by many in the vertebrate learning research community (and some in the invertebrate learning research community as well) precisely because the CS was found to elicit a response that resembled the CR (Halas, James &

Knutson, 1962; Halas, James & Stone, 1961).

Most of the widely accepted invertebrate models of classical conditioning do not conform to (satisfy) Gormezano and Kehoe's taxonomy. For example, two of the major invertebrate models of classical conditioning *(Limax* and *Pleurobranchaea)* use an Instrumental Approach Behavior design (i.e., taste aversion paradigm). Taste aversion learning might be better characterized as a case of punishment rather than as classical conditioning (Schwartz, 1974; Solomon, 1977). Another conditioning model *(Hermissenda)* may also be better characterized as the suppression of a taxic response rather than as classical conditioning (Crow & Alkon, 1978).

An improved *Hermissenda* training method (Lederhendler, Gart & Alkon, 1986) satisfies the criteria proposed by Gormezano and Kehoe to demonstrate classical conditioning, as does an eye withdrawal training technique developed for the green crab (Abramson & Feinman, 1988; Feinman, Llinas, Abramson & Forman, 1990). In both the improved *Hermissenda* training method and the eye withdrawal training technique developed for the green crab, the conditioned stimuli do not elicit a response resembling the conditioned response prior to CS-US pairings and the conditioned response (CR) emerges from the same effector system as the unconditioned response (UR).

The popular proboscis conditioning paradigm in which an olfactory stimulus (CS) is paired with a sucrose feeding (US) also meets the CS-CR relationship classified as classical conditioning by Gormezano and Kehoe. Proboscis extension in honeybees is perhaps the best behavioral model of classical conditioning in invertebrates (Menzel, Hammer, Braun, Mauelshagen & Sugawa, 1991). However, there are cases in which the data may be compromised by the presence of "spontaneous responders," animals that respond to the CS prior to any pairing with the US (Smith & Abramson, 1992). Moreover, the role of central excitatory state has seldom been accounted for in proboscis conditioning studies (Terry & Hirsch, this volume).

It is worth noting that Gormezano and Kehoe do not consider alpha conditioning an example of classical conditioning. Equating alpha conditioning with classical conditioning may inhibit any search for the evolutionary precursors of classical conditioning. If alpha

conditioning, best characterized as the association of two unconditioned stimuli, and classical conditioning, best characterized as the
association of a neutral stimulus with an unconditioned stimulus,
are considered identical then certain questions become impossible
to ask. For instance, which animal taxa were the earliest to associate
neutral events? Or, what are the behavioral profiles of nonneutral
and neutral stimuli? Although there are varying definitions of alpha
conditioning, it has been most precisely defined (Dyal & Corning,
1973; Razran, 1971) as a conditioned *associative* sensitization of an
innate response.

Another classical conditioning model *(Aplysia)* is considered
by some as a case of alpha conditioning (Farley & Alkon, 1985;
Schreurs, 1989). A distinction is often made in the conditioning
literature between alpha conditioning and classical conditioning.
Alpha conditioning refers to a response that is elicited by the CS
prior to training. A problem of interpretation occurs when the response elicited by the CS resembles the response one uses as an index of learning. Some members of the invertebrate learning community have forcefully argued that this is an artificial
distinction—that at the cellular level, alpha and "true" conditioned
responses may not represent separate processes but represent separate points on a response threshold continuum (Carew, Abrams,
Hawkins & Kandel, 1984). Other members of the research community, however, believe that alpha conditioning is not classical conditioning and that so-called classical conditioning in *Aplysia* is better characterized as a case of alpha conditioning (Farley & Alkon,
1985; Schreurs, 1989).

How is one to decide which conceptualization is the correct
one? The answer is that one must conduct a comparative analysis
within the framework of an agreed upon taxonomy.

Operant Conditioning. In contrast to studies of classical conditioning, and to studies of habituation and sensitization, there has been
limited application of the techniques of operant conditioning to invertebrates. As originally conceived, operant behavior is characterized by the "goal-directed" motor manipulation of the environment
(Lee, 1988). In place of the goal-directed modification of vertebrate
behavior that was the hallmark of the Skinnerian system, operant

conditioning of invertebrates generally consists of *any* behavior sensitive to response-reinforcer contingencies. Thus, operant conditioning of invertebrates has been considered to include such procedures as manipulating body position or a single appendage in response to aversive stimulation (Horridge, 1962); running against a taxic or kinetic preference, as, for example, when a cockroach is shocked upon entering a dark compartment (Szymanski, 1912), and learning various mazes and runways (Longo, 1964).

These procedures, however, might not constitute operant behavior. A major requirement of operant conditioning has been that species-typical behavior is minimized by interjecting a "novel" behavior, such as a lever press, or a nonarbitrary response brought under the control of a discriminative stimulus or cue placed between the animal and the animal's reception of some reward. In this way, the experimenter demonstrates that the animal has learned not only how to operate some device but also how to use it. Only a few invertebrate operant conditioning studies have attempted to shape the behavior of an invertebrate. Forman (Forman, 1984; Forman & Zill, 1984) trained locusts to receive a positive event or remove a negative one by maneuvering a leg to an arbitrarily selected position. Rubadeau and Conrad (1963) demonstrated food reinforced lever press behavior of the cockroach. A similar device has been used for the solitary bee (Pessotti, 1972). A preliminary report describing lever press performance for crayfish has also appeared (Olson & Strandberg, 1979). Lever press conditioning has been demonstrated in *Aplysia* (Downey & Jahan-Parwar, 1972), and, most recently, an unusual study of bar press by *Helix* has been described in which animals work for neuronal stimulation (Balaban & Chase, 1989).

Detailed behavioral analysis of operant conditioning of an invertebrate using the lever press technique has been developed for the green crab. Animals are rewarded for a lever press with a bit of squid. It has been demonstrated that each lever press does not need to be reinforced and that the animals can learn to press one of two levers in a spatial discrimination task (Abramson & Feinman, 1990a, 1990b). It has been reported also that at least some crabs can perform when the ratio of lever presses to reinforcement is 9 (i.e., fixed ratio of 9) (Feinman, Korthals-Altes, Kingson, Abramson & Forman, 1990). Some research (Abramson, unpublished observations) suggests that green crabs can not time their responses to arbitrarily selected durations, however. Further studies are required to learn whether the crustacean lever response is sensitive to omission of contingencies and whether some property of the lever press response class is sensitive to response-reinforcer contingencies. Until such data are available, operant behavior as a characterisitic ability of this crustacean must be considered tentative.

One might be more confident that an invertebrate technique is measuring an operant if it can be shown that (1) the operant response minimizes species-typical behavior, (2) some property of the response class such as its rate, force or interresponse time can be trained, (3) the response no longer occurs when such responses postpone the delivery of reward, and (4) the response can be brought under the control of a cue. Generally, the analysis of invertebrate operant behavior has not been carried very far. To my knowledge no one has successfully trained an invertebrate to adjust its running speed or press a lever with a degree of force set by the experimenter. It is not known whether any of the invertebrates, with the exception of the crab, can maintain responding on a schedule as simple as a fixed ratio of 2. Moreover, there are no data on the stability of the lever press response over time. Given the importance of understanding the neuronal and structural mechanisms of operant behavior, the lack of behavioral data in these operant paradigms is surprising (Carew & Sahley, 1986).

The problem of what constitutes operant behavior and its measurement must be solved before a meaningful comparative or simple system analysis of operant behavior can be undertaken in invertebrates.

Differences Between Avoidance and Punishment. Many studies contain the pervasive conceptual error of failing to distinguish between the paradigm of punishment and the paradigm of avoidance (Abramson, 1994). In punishment training, emitting a target response results in the *presentation* of the aversive event. However, in avoidance training, emitting the target response results in the *postponement* or *omission* of the aversive event. The failure to recognize this distinction stems from a misunderstanding of an early theory of punishment that stressed the acquisition of competing responses, that is, the passive avoidance theory (Dinsmore, 1954). This theory suggests that punishment is effective because stimuli associated with aversive reinforcers elicit behavior that competes with the target response. An animal engaged in such behavior "avoids" the punishment contingency by withholding the target response. From this perspective, cockroaches (Pritchatt, 1968, 1970) and fruit flies (Booker & Quinn, 1981) reportedly "avoid" shock in the well-known leg-position preparation, cockroaches "avoid" darkness when light is present (Minami & Dallenbach, 1946; Szymanski, 1912) and ants "avoid" an area associated with X-rays (Martinsen & Kimeldorf, 1972). In each of these experiments, a response results in the *presentation* of the punishing event. After a number of such response-reinforcer pairings the probability of response—whether it be the manipulation of a flexor or a more molar behavior—declines.

When signals are introduced into the punishment paradigm as controls for nonassociative effects, the difference between avoidance and punishment is evident. Discrimination among stimuli is but one of a number of traditional controls for the effect of stimulation *per se* in conditioning experiments. A widely used technique for flies, intended primarily to provide a target behavior for genetic analysis (Dudai, 1977; Hewitt, Fulker & Hewitt, 1983; Quinn et al., 1974), is considered by its advocates to represent signaled avoidance. In this mass conditioning situation, fruit flies are trained with two signals, one of which is associated with an aversive stimulus. Because presentation of the aversive stimulus is *contingent* upon the subject *entering* a particular portion of the apparatus, it can be interpreted, instead, as a punishment procedure with differential control.

In contrast to punishment experiments, the number of invertebrate signaled avoidance experiments of the kind used to study vertebrates is quite small. Signaled avoidance, in which the presentation of the CS leads to an *increase in the probability* of a target response, has been demonstrated in honeybees (Abramson, 1986; Smith, Abramson & Tobin, 1991) and green crabs (Abramson, Armstrong, Feinman & Feinman, 1988).

Signaled avoidance has been less convincingly demonstrated, because of lack of control procedures, in cockroaches (Chen, Aranda & Luco, 1970), earthworms (Ray, 1968), planarians (Ragland & Ragland, 1965) and crayfish (Taylor, 1971). The cockroach and planarian experiments have the additional complication that the avoidance signals were not neutral. Shock was used as a CS in the cockroach experiment, a questionable procedure, and light was used in the planarian work. Signaled avoidance has not been found in houseflies (Leeming, 1985), cockroaches (Pritchatt, 1970), horseshoe crabs (Makous, 1969) and earthworms (Kirk & Thompson, 1967). Unsignaled avoidance has been demonstrated in bees (Abramson, 1986) and cockroaches (Longo, 1964). In contrast with vertebrate studies, however, the invertebrate unsignaled data does not reveal the formation of temporal regularities so often characteristic of vertebrate unsignaled avoidance.

Taxonomies of Invertebrate Learning Paradigms Should Be Developed

Invertebrate learning literature needs to construct a classification schema or taxonomy to organize the various procedures used to measure nonassociative and associative learning in invertebrates. Many different procedures have been developed to measure invertebrate learning, which makes it difficult to determine that these procedures are measuring the same process. A taxonomy or classification based on precise manipulations of such training variables as stimulus intensity, interstimulus interval and pattern of reinforcement is needed to measure invertebrate learning. In the absence of such a classification, the various procedures designed to measure the learning of different invertebrate groups cannot be described as equivalent—or different.

It is interesting to note how much effort was devoted by Schneirla to just such parametric experiments. Even a casual browse through Maier and Schneirla (1935/1964) reveals such section headings as "Comparison of the maze with the discrimination box," "Learning situations as measures of learning," "A comparison of the conditioned response with other forms of learning," "Quantitative differences in mazes and their effects on learning," "The effect of different types of junctions on maze learning," "The effect of the maze pattern upon behavior in the maze," "The kind of blind alley used and its effect on behavior," to mention just a few. Such attention to parametric variables is missing from contemporary simple systems work.

Several taxonomies have been proposed: Dyal and Corning (1973); Gormezano and Kehoe (1975) for classical conditioning; Woods (1974) for instrumental and operant conditioning. Woods' (1974) classification of instrumental conditioning identifies 16 categories of conditioning based on the presence or absence of a discriminative stimulus and the desirability of the reward.

Tulving (1985) describes six ways in which a classification scheme can advance the field of learning and memory. First, a taxonomy would give some theoretical structure to the design and analysis of experiments. Presently, it is difficult to find a central theme in the invertebrate learning literature. Studies of invertebrate simple system preparations appear to come and go with the only rationale for developing a preparation seeming to be the exploitation of a nervous system. Second, a taxonomy would replace the use of general categories such as "classical conditioning" and "operant conditioning" with detailed descriptors of the procedures. This would assist in the ability to generalize empirical facts across invertebrate species and preparations. Third, theoretical processes evoking, for instance, neuronal and molecular models of invertebrate behavior could be specified with greater precision. We will no longer risk the possibility of not knowing what type of behavior our neuronal and molecular models represent. Fourth, novel procedures and results can be described easily in terms of the amount of deviation from specified categories. Fifth, a classification system would stimulate a comparative approach to the study of learning and memory. It would provide, for instance, direction for conducting parametric analysis.

The sixth way in which a classification system can advance the field of learning and memory would be to enable those scientists interested in behavioral problems to speak the same "language." Confusion over the differences between, for instance, punishment and avoidance, operant and instrumental conditioning, and the various types of classical conditioning would be eliminated. As Bitterman (1962) noted over 30 years ago, "Classification is not merely a matter of taste" (p. 81). Without an objective strategy to classify invertebrate conditioning techniques, neuronal and molecular models of conditioning, while certainly important, will forever rest upon a weak foundation of behavioral knowledge.

Let us consider how the lack of a generally accepted taxonomy presents problems for the classical conditioning of *Aplysia*. One example of this was discussed in regard to alpha conditioning.

Aplysia conditioning phenomena picture can be complicated further by making the seemingly plausible assumption that a conditioned response to electric shock is formed because the withdrawal response (the CR in such experiments) reduces the amount of electric current flowing through the animal (to my knowledge no one has measured the amount of current flowing through a contracted *Aplysia*). Rats placed in similar conditions often are observed to "fold up" to stimuli predicting shock. Such behavior on the part of the rat reduces the amount of current flowing through it and therefore, reduces the amount of physical discomfort (Campbell & Masterson, 1969). If this assumption is confirmed in *Aplysia* by measuring the amount of current flowing through the animal during a withdrawal conditioned response, then alpha conditioning in *Aplysia* would formally fit the definition of instrumental training offered by Razran (1971). In our example, the reinforcement would be shock reduction, and the preexisting response would be the slight withdrawal response elicited by the CS prior to any pairing with the US.

The *Aplysia* physiological and biochemical models could, therefore, just as easily be applied to instrumental conditioning as to classical conditioning. How is one to decide which behavioral model to use? The answer once again is that one must place a greater emphasis on conducting a comparative analysis within the framework of a taxonomy.

The Reporting of Individual Level Data, as Contrasted with Species or Other Population Classifications Should Be Encouraged

In contrast to vertebrate work particularly in operant studies, we find few reports of individual data in invertebrate learning research (Abramson, 1994). In some cases, as in honeybee classical conditioning data, the data are often presented as the percentage of animals responding on each trial. In other cases, such as *Aplysia* classical conditioning, the data are presented as group means. Grouped data do not give the shape of individual learning curves nor information about the variation among animals. Moreover, the number of records discarded from a test population are rarely reported. Without such data it is difficult to know how many animals from a given population do indeed learn. Thus, the reliance on group data could lead to statements about species characteristics that are not reliable or valid. A similar criticism was made by Hirsch and Holliday (1988) to the fruit fly mass conditioning situation developed by Quinn (Quinn et al., 1974) and subsequently used by various experimenters.

It is interesting to compare the lack of individual data in contemporary studies of invertebrate learning with those provided by Schneirla in his famous studies of ant learning (e.g., Schneirla, 1929, 1941, 1943). Indeed it is difficult to find an empirical paper of his that does contain at least one example of individual data (Aronson, Tobach, Rosenblatt & Lehrman, 1972). Little individual data are published for the classical conditioning models of *Limax, Aplysia, Apis* and *Hermissenda*. Some exceptions are: the Greenberg, Castellucci, Bayley and Schwartz (1987) study with *Aplysia* and the proboscus extension paradigm in honeybees (Smith et al., 1991). In the *Aplysia* study, the caption for Figure 1 gives the high degree of variability encountered in behavioral experiments with this organism. This is the only reference I have found to individual variability in the *Aplysia* model. The Smith et al. paper (1991) reveals that the group learning curve is not always representative of individual honeybee performance.

There are few reports of individual data in operant conditioning investigations as well. Cumulative records of individual performance are available only for the crab (Abramson & Feinman, 1990b) and bee (Grossmann, 1973; Sigurdson, 1981). More often than not, when individual operant data are available, they are in the form of chart recordings, rather than cumulative records, and suggest long pauses between responses rather than the smooth steady rate of responding associated with vertebrate operant conditioning (Balaban & Chase, 1989). I would suggest that at least in the initial demonstration of operant conditioning, the data be presented in the form of cumulative records. This would obviously facilitate comparisons between vertebrates and invertebrates and give some idea of the individual variability of behavior. A national database of individual learning data would be extremely useful for those studying invertebrate learning and those interested in mathematical modeling of behavior. Many master and doctoral theses contain such raw data and might contribute to that database (Hirsch, personal communication).

Cognitive Explanations of Invertebrate Learning Phenomena Presents a Problematic Issue

The trend toward interpreting invertebrate learning in terms of "representations," "cognitive maps" and other language borrowed from the vocabulary of human information processing may be unwarranted at this time. For instance, a recent report interprets the simple choice behavior of a bumblebee as using "computational rules" and "cognitive architectures" (Real, 1991). Lloyd (1986) prefers to interpret the modification of tropisms in *Hermissenda* as the manipulation of the animals' "inner representation." The concept of honeybee "language" is another example of this trend. Although not generally remembered, von Frisch consistently placed quotations marks around the word "language" in his latest writings.

Before beginning the story I should like to emphasize the limitations of the language metaphor. The true comparative linguist is concerned with one of the subtlest products of man's powerfully developed thought processes. The brain of a bee is the size of a grass seed and is not made for thinking. The actions of bees are mainly governed by instinct. Therefore the student of even so complicated and purposeful an activity as the communication dance must remember that he is dealing with innate patterns, impressed on the nervous system of the insects over the immense reaches of time of their phylogenetic development. (von Frisch, 1962 p. 79; see also Schneirla, 1962)

Unfortunately, von Frisch's metaphor of bee "language" has been so carelessly paraphrased over the years that a generation of scientists, students and nonspecialists seem convinced that he demonstrated that bees possess a symbolic form of communication.

Another example of the extension of cognitive concepts to invertebrate learning is represented by the effect of contemporary views of classical conditioning, based on vertebrates, on our thinking about conditioning in invertebrates. For Pavlov and his contemporaries, classical conditioning involved the straightforward associations of two events occurring closely in time. As a result of such vertebrate experiments as blocking, overshadowing and manipulating the contingency between the CS and US, higher vertebrates are now thought to detect and evaluate information embedded in the classical conditioning situation (Mackintosh, 1983). Some investigations of invertebrate learning studies have replicated these experiments with invertebrates (using CSs that are not neutral) and suggest that their preparations can tap into the analogous vertebrate processes (Farley, 1987; Hawkins, Carew & Kandel, 1986; Sahley, Rudy & Gelperin, 1981).

The trend to characterize vertebrate learning in terms of cognitive concepts is not without challenge (Amsel, 1989). Even if it is concluded that some vertebrate species possess cognitive structures, it remains doubtful whether invertebrates also possess such structures. Blocking, overshadowing and successive negative contrast in honeybees, for example, can be understood without recourse to cognitive constructs (Couvillon & Bitterman, 1982, 1984). When attempting to interpret invertebrate behavior in terms of cognitive constructs, it is important to recall C. Lloyd Morgan's canon: "In no case may we interpret an action as the outcome of the exercise of a higher psychical faculty, if it can be interpreted as the outcome of the exercise of one which stands lower in the psychological scale" (Morgan, 1894/1977, p. 53).

Summary

It has been the purpose of this chapter to point out what I feel are inconsistencies in the invertebrate learning literature, especially as it applies to simple systems work. The invertebrate learning literature conducted within a framework of simple systems research suggest that invertebrates can readily be conditioned using a number of nonassociative and associative paradigms. However, several limitations of the existing data are readily apparent. Although there are literally hundreds of studies examining learning in invertebrates, there is no generally accepted taxonomy of the procedures used to generate learning. Classical and operant conditioning have been demonstrated in various invertebrates, but there is a lack of parametric work that would enable researchers to ensure that the various procedures used to generate learning are indeed producing the same result.

Furthermore, the invertebrate learning literature suggests that terms used to describe learning phenomena are not being applied consistently. For example, what I might cite as an instance of punishment, another researcher may call avoidance. What another might call a conditioned stimulus, I might call an unconditioned stimulus.

In addition to the lack of an accepted taxonomy of learning and the lack of consistency in the application of terms used to describe learning phenomena, the majority of invertebrate studies report only group data. Thus, much information is available about learning in a generalized social insect, crustacean and mollusc, but very little information is available as to whether individual invertebrates behave in a similar way.

Additional problems can be found in the trend to stress the similarities of learning in invertebrates and vertebrates without noting the differences, and the use of cognitive concepts to explain invertebrate learning in the absence of supporting data. The title of this chapter, "Where have I heard it all before?" can be found in the writings of T. C. Schneirla. In fact one of the more depressing aspects of the invertebrate learning literature is that these issues never seem to get resolved. They are passed on from one generation of behavioral scientist to the next.

Acknowledgments

A version of this chapter has appeared in *A primer of invertebrate learning: The behavioral perspective*. The author would like to thank Dr. Ethel Tobach and Dr. A. J. Figueredo for their comments, suggestions and criticisms.

References

Abramson, C. I. (1986). Aversive conditioning in honeybees *(Apis mellifera)*. *Journal of Comparative Psychology, 100*, 108–116.

Abramson, C. I. (1994). *A primer of invertebrate learning: The behavioral perspective*. Washington, D.C., American Psychological Association.

Abramson, C. I., Armstrong, P. M., Feinman, R. A., & Feinman, R. D. (1988). Signaled avoidance learning in the eye withdrawal reflex of the green crab. *Journal of the Experimental Analysis of Behavior, 50*, 483–492.

Abramson, C. I., & Feinman, R. D. (1988). Classical conditioning of the eye withdrawal reflex in the green crab. *Journal of Neuroscience, 8*, 2907–2912.

Abramson, C. I., & Feinman, R. D. (1990a). Operant conditioning in crustaceans. In K. Wiese, J. Tautz, H. Reichert & B. Mulloney (eds.), *Frontiers in crustacean neurobiology* (pp. 207–214). Berlin: Birkhauser-Verlag.

Abramson, C. I., & Feinman, R. D. (1990b). Lever-press conditioning in the crab. *Physiology & Behavior, 48*, 267–272.

Akahane, R., & Amakawa, T. (1983). Stable and unstable phase of memory in classically conditioned fly, *Phormia regina*: Effects of nitrogen gas anaesthesia and cycloheximide injection. *Journal of Insect Physiology, 29*, 331–337.

Amsel, A. (1989). *Behaviorism, neobehaviorism and cognitivism in learning theory: Historical and contemporary perspectives*. Hillsdale, N.J.: Lawrence Erlbaum Associates.

Aronson, L. R., Tobach, E., Rosenblatt, J. S., & Lehrman, D. S. (1972). *Selected writings of T. C. Schneirla*. San Francisco: W. H. Freeman.

Balaban, P. M., & Chase, R. (1989). Self-stimulation in snails. *Neuroscience Research Communications, 4,* 139–146.

Beach, F. A. (1950). The snark was a boojum. *American Psychologist, 5,* 115–124.

Bitterman, M. E. (1962). Techniques for the study of learning in animals: Analysis and classification. *Psychological Bulletin, 59,* 81–93.

Bitterman, M. E. (1975). The comparative analysis of learning. *Science, 188,* 699–709.

Bitterman, M. E. (1988). Vertebrate-invertebrate comparisons. In H.J. Jerison & I. Jerison (eds.), *Intelligence and evolutionary biology* (pp. 251–276). Berlin: Springer-Verlag.

Bitterman, M. E., Menzel, R., Fietz, A., & Schäfer, S. (1983). Classical conditioning of proboscis extension in honeybees *(Apis mellifera). Journal of Comparative Psychology, 97,* 107–119.

Bolles, R. C. (1983). The explanation of behavior. *Psychological Record, 33,* 31–48.

Booker, R., & Quinn, W. G. (1981). Conditioning of leg position in normal and mutant *Drosophila. Proceedings of the National Academy of Science, 78,* 3940–3944.

Brookshire, K. H. (1970). Comparative psychology of learning. In M. H. Marx (ed.), *Learning: Interactions* (pp. 291–347). London: Macmillan.

Bullock, T. H. (1966). Simple systems for the study of learning mechanisms. *Neuroscience Research Program Bulletin, 4,* 104–233.

Campbell, B. A., & Masterson, F. A. (1969). Psychophysics of punishment. In B. A. Campbell & R. M. Church (eds.), *Punishment and aversive behavior* (pp. 3–42). New York: Appleton-Century-Crofts.

Carew, T. J., Abrams, T. W., Hawkins, R. D., & Kandel, E. R. (1984). The use of simple invertebrate systems to explore psychological issues related to associative learning. In D. L. Alkon & J. Farley (eds.), *Primary neural substrates of learning and behavioral change* (pp. 169–183). Cambridge: Cambridge University Press.

Carew, T. J., & Sahley, C. L. (1986). Invertebrate learning and memory: From behavior to molecules. *Annual Review of Neuroscience, 9,* 435–487.

Carew, T. J., Walters, E. T., & Kandel, E. R. (1981). Classical conditioning in a simple withdrawal reflex in *Aplysia californica. Journal of Neuroscience, 1,* 1426–1437.

Catania, A. C. (1973). The concept of the operant in the analysis of behavior. *Behaviorism, 1,* 103–116.

Chen, W. Y., Aranda, L. C., & Luco, J. V. (1970). Learning and long- and short-term memory in cockroaches. *Animal Behaviour, 18,* 725–732.

Couvillon, P. A., & Bitterman, M. E. (1982). Compound conditioning in honeybees. *Journal of Comparative and Physiological Psychology, 96,* 192–199.

Couvillon, P. A., & Bitterman, M. E. (1984). The overlearning-extinction effect and successive negative contrast effect in honeybees *(Apis mellifera). Journal of Comparative Psychology, 98,* 100–109.

Crow, T. J., & Alkon, D. L. (1978). Retention of an associative behavioral change in *Hermissenda. Science, 201,* 1239–1241.

Dinsmore, J. A. (1954). Punishment I. The avoidance hypothesis. *Psychological Review, 61,* 34–46.

Downey, P., & Jahan-Parwar, B. (1972). Cooling as reinforcing stimulus in *Aplysia. American Zoologist, 12,* 507–512.

Dudai, Y. (1977). Properties of learning and memory in *Drosophila melanogaster. Journal of Comparative Physiology, 114,* 69–89.

Dunn, P. D. C., & Barnes, W. J. P. (1981). Learning of leg position in the shore crab, *Carcinus maenas. Marine Biology, 8,* 67–82.

Dyal, J. A., & Corning, W. C. (1973). Invertebrate learning and behavior taxonomies. In W. C. Corning, J. A. Dyal & A. O. D. Willows (eds.), *Invertebrate learning, Vol.1. Protozoans through annelids* (pp. 1–48). New York: Plenum.

Farley, J. (1987). Contingency learning and causal detection in *Hermissenda*: I. Behavior. *Behavioral Neuroscience, 101,* 13–27

Farley, J., & Alkon, D. L. (1985), Cellular mechanisms of learning, memory and information storage. *Annual Review of Psychology, 36,* 419–494.

Feinman, R. D., Korthals-Altes, H., Kingson, S., Abramson, C. I., & Forman, R. R. (1990). Lever-press conditioning in the crab. Green crabs perform well on fixed ratio schedules but can they count? *Biological Bulletin 179*, 233 (abstract).

Feinman, R. D., Llinas, R. H., Abramson, C. I. & Forman, R. R. (1990). Electromyographic record of classical conditioning of eye withdrawal in the crab. *Biological Bulletin, 178*, 187–194.

Forman, R. R. (1984). Leg position learning by an insect. I. A heat avoidance learning paradigm. *Journal of Neurobiology, 15*, 127–140.

Forman, R. R., & Zill, S. (1984). Leg position learning by an insect. II. Motor strategies underlying learned leg extension. *Journal of Neurobiology, 15*, 221–237.

Frisch, K. von. (1962). Dialects in the language of the bees. *Scientific American, 207*, 79–87.

Fukushi, T. (1979). Properties of olfactory conditioning in the housefly, *Musca domestica. Journal of Insect Physiology, 25*, 155–159.

Gormezano, I. (1984). The study of associative learning with CS-CR paradigms. In D. L. Alkon & J. Farley (eds). *Primary neural substrates of learning and behavioral change* (pp. 5–24). Cambridge: Cambridge University Press.

Gormezano, I., & Kehoe, E. J. (1975). Classical conditioning: Some methodological-conceptual issues. In W. K. Estes (ed.), *Handbook of learning and cognitive processes: Vol. 2. Conditioning and behavior theory* (pp. 143–179). Hillsdale, N.J.: Lawrence Erlbaum Associates.

Gormezano, I., Kehoe, E. J., & Marshall, B. S. (1983). Twenty years of classical conditioning research with the rabbit. In J. M. Sprague & A. N. Epstein (eds.). *Progress in physiological psychology* (pp. 197–275). New York: Academic Press.

Greenberg, S. M., Castellucci, V. F., Bayley, H., & Schwartz, J. H. (1987). A molecular mechanism for long term sensitization in *Aplysia. Nature, 329*, 62–65.

Grossmann, K. E. (1973). Continuous, fixed-ratio and fixed-interval reinforcement in honey bees. *Journal of the Experimental Analysis of Behavior, 20*, 105–109.

Halas, E. S., James, R. L., & Knutson, C. (1962). An attempt at classical conditioning in the planarian. *Journal of Comparative and Physiological Psychology, 55*, 969–971.

Halas, E. S., James, R. L., & Stone, L. A. (1961). Types of responses elicited in planaria by light. *Journal of Comparative and Physiological Psychology, 54*, 302–305.

Hawkins, R. D., Carew, T. J., & Kandel, E. R. (1986). Effects of interstimulus interval and contingency on classical conditioning of the *Aplysia* siphon withdrawal reflex. *Journal of Neuroscience, 6*, 1695–1701.

Hewitt, J. K., Fulker, D. W., & Hewitt, C. A. (1983). Genetic architecture of olfactory discriminative avoidance conditioning in *Drosophila melanogaster. Journal of Comparative Physiology, 97*, 52–58.

Hirsch, J., & Holliday, M. (1988). A fundamental distinction in the analysis and interpretation of behavior. *Journal of Comparative Psychology, 102*, 372–377.

Horridge, G. A. (1962). Learning of leg position by headless insects. *Nature, 193*, 697–698.

Hull, C. L. (1943). *Principles of behavior.* New York: Appleton-Century-Crofts.

Kirk, W. E., & Thompson, R. W. (1967). Effects of light, shock, and goal box conditions on runway performance of the earthworm, *Lumbricus terrestris. Psychological Record, 17*, 49–54.

Lederhendler, I. I., Gart, S., & Alkon, D. L. (1986). Classical conditioning of *Hermissenda:* Origin of a new response. *Journal of Neuroscience, 6*, 1325–1331.

Lee, V. L. (1988). *Beyond behaviorism.* Hillsdale, N.J.: Lawrence Erlbaum Associates.

Leeming, F. C. (1985). Free-response escape but not avoidance learning in houseflies *(Musca domestica). Psychological Record, 35*, 513–523.

Lejeune, H., & Wearden, J. H. (1991). The comparative psychology of fixed-interval responding: Some quantitative analysis. *Learning and Motivation, 22*, 84–111.

Lloyd, D. (1986). The limits of cognitive liberalism. *Behaviorism, 14*, 1–14.

Longo, N. (1964). Probability learning and habit reversal in the cockroach. *American*

Journal of Psychology, 77, 29–41.

Mackintosh, N. J. (1983). *Conditioning and associative learning*. Oxford: Clarendon Press.

Macphail, E. M. (1987). The comparative psychology of intelligence. *Behavioral and Brain Sciences, 10*, 645–695.

Maier, N. R. F., & Schneirla, T. C. (1964). *Principle of animal psychology*. Revised ed. New York: Dover. (Original work, 1935).

Makous, W. L. (1969). Conditioning in the horseshoe crab. *Psychonomic Science, 14*, 406.

Martinsen, D. L., & Kimeldorf, D. J. (1972). Conditioned spatial avoidance behavior of ants induced by X-rays. *Psychological Record, 22*, 225–232.

McGuire, T. R. (1984). Learning in three species of diptera: The blow fly *Phormia regina*, the fruit fly *Drosophila melanogaster*, and the house fly *Musca domestica*. *Behavior Genetics, 14*, 479–526.

Menzel, R., Hammer, M., Braun, G., Mauelshagen, J., & Sugawa, M. (1991). Neurobiology of learning and memory in honeybees. In L. J. Goodman & R. C. Fisher (eds.), *The behavior and physiology of bees* (pp. 323–353). London: C. A. B. International.

Minami, H., & Dallenbach, K. M. (1946). The effect of activity upon learning and retention in the cockroach. *American Journal of Psychology, 59*, 1–58.

Morgan, C. L. (1894). *Introduction to comparative psychology*. London: Walter Scott. Reprinted in D. N. Robinson (ed.) 1977. *Significant contributions to the history of psychology: Series D, Volume 2, C. L. Morgan* (pp. 1–382). Washington: University Publications of America.

Olson, G. S., & Strandberg, R. (1979). Instrumental conditioning in crayfish: Lever pulling for food. *Society for Neuroscience Abstracts, 5*, 257.

Pessotti, I. (1972). Discrimination with light stimuli and a lever-pressing response in *Melipona rufiventris*. *Journal of Apicultural Research, 11*, 89–93.

Pritchatt, D. (1968). Avoidance of electric shock by the cockroach *Periplaneta americana*. *Animal Behaviour, 16*, 178–185.

Pritchatt, D. (1970). Further studies on the avoidance behaviour of *Periplaneta americana*. *Animal Behaviour, 18*, 485–492.

Quinn, W. G., Harris, W. A., & Benzer, S. (1974). Conditioned behavior in *Drosophila melanogaster*. *Proceedings of the National Academy of Science, 71*, 708–712.

Ragland, R. S., & Ragland, J. B. (1965). Planaria: Interspecific transfer of a conditionability factor through cannibalism. *Psychonomic Science, 3*, 117–118.

Ray, A. J. (1968). Instrumental light avoidance by the earthworm. *Communications in Behavioral Biology, 1*, 205–208.

Razran, G. (1971). *Mind in evolution: An east-west synthesis of learned behavior and cognition*. Boston: Houghton Mifflin.

Real, L. A. (1991). Animal choice behavior and the evolution of cognitive architecture. *Science, 253*, 980–986.

Rubadeau, D. O., & Conrad, K. A. (1963). An apparatus to demonstrate and measure operant behavior of arthropoda. *Journal of the Experimental Analysis of Behavior, 6*, 429–430.

Sahley, C. L. (1984). Behavior theory and invertebrate learning. In P. Bateson & P. Marler (eds.), *Biology of learning* (pp. 181–196). Berlin: Springer-Verlag.

Sahley, C., Rudy, J. W., & Gelperin, A. (1981). An analysis of associative learning in a terrestrial mollusc. I. Higher-order conditioning, blocking, and a transient US preexposure effect. *Journal of Comparative Physiology, 144*, 1–8.

Schick, K. (1971). Operants. *Journal of the Experimental Analysis of Behavior, 15*, 413–423.

Schneirla, T. C. (1929). Learning and orientation in ants. *Comparative Psychology Monographs, 6*, 1–143.

Schneirla, T. C. (1941). Studies on the nature of ant learning: I. The characteristics of a distinctive initial period of generalized learning. *Journal of Comparative Psychology, 32*, 41–82.

Schneirla, T. C. (1943). The nature of ant learning: II. The intermediate stage of segmental maze adjustment. *Journal of Comparative Psychology, 34,* 149–176.

Schneirla, T. C. (1949). Levels in the psychological capacities of animals. In R. W. Sellars, V. J. McGill & M. Farber (eds.), *Philosophy for the future: The quest of modern materialism* (pp. 243–286). New York: Macmillan.

Schneirla, T. C. (1962). Psychology, Comparative. *Encyclopaedia Britannica,* Vol. 18 (pp. 690Q–703). Reprinted in L. R. Aronson, E. Tobach, J. S. Rosenblatt & D. S. Lehrman (eds.) (1972), *Selected writings of T. C. Schneirla* (pp. 30–85). San Francisco: W. H. Freeman.

Schoenfeld, W. N. (1976). The "response" in behavior theory. *Pavlovian Journal of Biological Science, 11,* 129–149.

Schreurs, B. G. (1989). Classical conditioning of model systems: A behavioral review. *Psychobiology, 17,* 145–155.

Schwartz, B. (1974). On going back to nature: A review of Seligman and Hager's Biological boundaries of learning. *Journal of the Experimental Analysis of Behavior, 21,* 183–198.

Sidowski, J. B. (1966). *Experimental methods and instrumentation in psychology.* New York: McGraw-Hill.

Sigurdson, J. E. (1981). Automated discrete trials techniques of appetitive conditioning in honeybees. *Behavior Research Methods & Instrumentation, 13,* 1–10.

Smith, B. H. & Abramson, C. I. (1992). Insect learning: Case studies in comparative psychology. In L .I. Nadel & J. Byrne (eds.), *Encyclopedia of learning and memory* (pp. 276–283). New York: Macmillian.

Smith, B. H., Abramson, C. I., & Tobin, T .R. (1991). Conditional withholding of proboscis extension in honeybees *(Apis mellifera)* during discriminative punishment. *Journal of Comparative Psychology, 105,* 345–356.

Solomon, R. L. (1977). An opponent-process theory of motivation: V. Affective dynamics of eating. In L. M. Barker, M. R. Best & M. Domjan (eds.), *Learning mechanisms in food selection* (pp. 255–269). Waco, Tex.: Baylor University Press.

Squire, L. R. (ed.) (1992). *Encyclopedia of learning and memory.* New York: Macmillan.

Staddon, J. E. R., & Bueno, J. L. O. (1991). On models, behaviorism and the neural basis of learning. *Psychological Science, 2,* 3–11.

Staddon, J. E. R., & Zhang, Y. (1989). Response selection in operant learning. *Behavioral Processes, 20,* 189–197.

Szymanski, J. S. (1912). Modification of the innate behavior of cockroaches. *Journal of Animal Behavior, 2,* 81–90.

Tavolga, W. N. (1969). *Principles of animal behavior.* New York: Harper & Row.

Taylor, R. C. (1971). Instrumental conditioning and avoidance behavior in the crayfish. *Journal of Biological Psychology, 13,* 36–41.

Teyler, T. J., Baum, W. M., & Patterson, M. M. (ed.) (1975). Behavioral and biological issues in the learning paradigm. *Physiological Psychology, 3,* 65–72.

Thompson, T. & Zeiler, M. D. (1986). *Analysis and integration of behavioral units.* Hillsdale, N.J.: Lawrence Erlbaum Associates.

Thompson, R., & McConnell, J. V. (1955). Classical conditioning in the planarian, *Dugesia dorotocephala. Journal of Comparative and Physiological Psychology, 48,* 65–68.

Thorpe, W. H. (1956). *Learning and instinct in animals.* Cambridge: Harvard University Press.

Tully, T., & Quinn, W. G. (1985). Classical conditioning and retention in normal and mutant *Drosophila melanogaster. Journal of Comparative Physiology A, 157,* 263–277.

Tulving, E. (1985). On the classification problem in learning and memory. In L. Nilsson & T. Archer (eds.), *Perspectives on learning and memory* (pp. 67–94). Hillsdale, N.J.: Lawrence Erlbaum Associates.

Warden, C. J., Jenkins, T. N., & Warner, L. H. (1940). Comparative psychology, a comprehensive treatise: Vol. 2. Plants and invertebrates. New York: Ronald Press.

Washburn, M. F. (1908). *The animal mind: A text-book of comparative psychology.*
    New York: Macmillan.
Willner, P. (1978). What does the headless cockroach remember? *Animal Learning &
    Behavior, 6,* 249–257.
Woods, P. J. (1974). A taxonomy of instrumental conditioning. *American Psychologist,
    29,* 584–596.
Zaldivar, M. E., & Jaffé, K. (1987). An automated training system for crickets. *Acta
    Cientifica Venezolana, 38,* 122–125.

Macroevolution of a Fixed Action Pattern for Learning

The Exploration Flights of Bees and Wasps

Rudolf Jander

Exploration is the learning behavior by which topographic knowledge is acquired. Topographic knowledge is spatial knowledge about the topography of one's home range, to be used in topographic or home-range orientation (Jander, 1975). Topographic orientation is well developed in some mollusks, many arthropods and most, if not all, vertebrates. Among arthropods, topographic orientation is best studied in the Aculeate Hymenoptera, the wasps, bees and ants (hundreds of publications, reviewed by Wehner, 1992), and among the vertebrates in the rodent families Muridae and Cricetidae (thousands of publications, reviewed by Poucet, 1993).

Phylogenetic analysis tells us that the last common ancestor between the arthropods and vertebrates dates back more than half a billion years and is inferred to have had a level of behavioral and morphological organization not more complex than that of extant planarians (Platyhelminthes) or of ribbon worms (Nemertea) (Arbas et al., 1991; Bowring et al., 1993; Miklos, 1993; Morris, 1993). Hence, it is safe to say that the mechanisms of topographic orientation in Hymenoptera and Rodentia have evolved independently at the supracellular level: sensory and motor structures linked by neuronal circuits. Indeed, there is not the slightest evidence for any supracellular homologies between the nervous systems and sensory organs of arthropods and vertebrates (Bitterman & Couvillon, 1991).

One would think that the remarkably complex analogy (homoplasy) between the well-studied topographic orientation systems of the Hymenoptera and of the rodents had been extensively compared. How wrong! Neither of the two major recent reviews on animal orientation covers the topographic orientation of the rodents, only that of the Hymenoptera (Schöne, 1984; Papi, 1992). For the only pertinent crossphyletic comparison one has to hark back all the way to Schneirla's limited comparison of maze learning and orientation in ants and rodents (Schneirla, 1941, 1943).

In this chapter, I will continue where Schneirla left off half a century ago. After a brief comparison of exploration in rodents and Hymenoptera, I will first describe and then reconstruct the function and macroevolution of the Exploration Flight of the Hymenoptera based on published findings and my own yet unpublished observations and experiments (Jander, 1994). Finally, an attempt will be made to fit the hymenopteran Exploration Flight into the conceptual framework of contemporary learning theories.

Exploration in Rodents and Hymenoptera

Exploration and inspection are investigatory behaviors whose main adaptive function is the acquisition of knowledge that might be useful in the future. Investigatory learning differs from the type of learning traditionally labeled conditioning by not requiring explicit reinforcement, specifically, reward or punishment signaling instantaneous biological benefits or costs. It is mainly for this distinction that it has often been referred to as latent learning (Blodgett, 1929; Mackintosh, 1974; Honig, 1987). The underlying motivation is curiosity, for which the necessary incentive is novelty in the perceived environment. Novelty is a currently perceived experience that mismatches with any remembered experience of the past. Novel stimulation may gain additional strength as incentive for curiosity if it co-incites additional appetitive or aversive drives.

Examples for curiosity that is co-incited in this way together with other drives and their associated actions are known for both vertebrates and Hymenoptera. A case for rodents is the rat's (*Rattus norvegicus*) investigatory (inspection and exploration) behavior after stimulation with an unexpected or novel electric shock (Hudson, 1950; Keith-Lucas & Guttman, 1975; Russel, 1973). A case for Hymenoptera is the honey bee's Exploration Flight after exposure to rich novel food resources or a familiar food resource in a suddenly novel environment (Jander, 1990, 1994; Lehrer, 1991, 1993; Opfinger, 1931). Such postreinforcement exploratory learning in rodents and bees has sometimes been termed "backward condition" (e.g., Keith-Lucas & Guttman, 1975; Lehrer, 1993), because stimuli perceived after reinforcement are learned. However, it is important to bear in mind that there is no evidence that all the learning operations called "backward conditioning" share more than just very superficial similarities.

Exploration is a category of investigatory behavior to which also inspection belongs. Inspection serves the acquisition of knowledge about objects. The rich literature on the investigatory behavior of rodents and other vertebrates has repeatedly been reviewed (Berlyne, 1950, 1960, 1966; Fowler, 1965; Halliday, 1968; Barnett, 1975; Archer & Birke, 1983; Voss & Keller, 1983; Gibson, 1988; Thinus-Blanc et al., 1991). For the near-target exploration in the Aculeate Hymenoptera, this is the first synopsis.

Exploration in rodents and in nest-building Aculeate Hymenoptera shares a series of analogous similarities: In both taxa, exploration is invariably anchored to a natural or artificial home base, typically a burrow, nest or nesting site. If, upon leaving such a site, the undisturbed rodent, bee, wasp or ant encounters a novel environment, they all perform a characteristic locomotor pattern that involves repetitive excursions of increasing distances and varying directions. Schneirla (1941) described such exploratory movements for the formicine ant *Formica schauffusii*, and various authors did so for similar aerial exploration in honeybee workers, which were released from a box in a foreign environment (Bethe 1898; Buttel-Reepen, 1915; Wolf, 1926; Becker, 1958). Similar movement patterns during exploration have been described for a wide variety of vertebrates, especially for rodents (Bovet, 1992).

Structure and Behavioral Context of Aculeate Exploration Flight

Flying bees and wasps are easily observed only near their targets, the locations where they find their various resources, nests, flowers, hunting grounds, etc. This constrains our current understanding of their exploratory behavior largely to these small areas within home ranges that cover up to many square kilometers in larger species (Janzen, 1971; Seeley, 1985). When flying from one such target to another one, they most of the time make a beeline immediately after takeoff. Sometimes, however, immediately after takeoff, and before initiat-

ing the straight line departure, they perform near the target a complex flight maneuver, the Exploration Flight.

Bates (1863) was the first to observe such flights in the small
pale-green sphecid (Bembicini) wasp *Microbembex ciliata*: After fin-
ishing the excavation of their burrows, the wasps took flight, turned
a few circles and then departed straight. Bates correctly interpreted
this behavior as local exploration for the purpose of recording local
orientation cues that facilitate the later return to the new nest. After
Bates, numerous investigators gave more or less complete descriptions of such flights for paper wasps (Vespidae), digger wasps
(Sphecidae) and bees (Apiformes). Only a small selection of references may suffice. For the Vespidae: Freisling, 1943; Kemper &
Döhring, 1967; Akre et al., 1976. For the Sphecidae: Tinbergen,
1932; Baerends, 1941; Iersel & Assem, 1964; Evans, 1966;
Chmurzynski, 1967; Zeil, 1993a, b. For the Apoidea: Buttel-Reepen,
1915; Frison, 1930; Opfinger, 1931; Plath, 1934; Michener, 1953,
1966; Manning, 1956; Becker, 1958; Frisch, 1967; Münster-
Swendsen, 1968; Vollbehr, 1975; Rajotte, 1979; Klinksik &
Schönitzer, 1988; Wcislo, 1992; Lehrer, 1993. I have observed hundreds of instances of this Exploration Flight in dozens of species
belonging to the same three taxa.

In much of the entomological literature this Exploration Flight
is referred to as "orientation flight." I prefer the former term because
using different names for the same behavior in vertebrates and insects, together with making "orientation" a synonym of "exploration," is misleading and impedes theoretical integration.

Most uses of the Exploration Flight have been observed at nesting sites, some in bees (*Apis, Bombus*) at feeding sites (Opfinger,
1931; Manning, 1956; Lehrer, 1991, 1993). I have seen a worker of
Vespula maculifrons engage in an Exploration Flight upon leaving a
dead branch from which it had collected fibers for nest building.
Confirming Opfinger (1931) and Manning (1956), I saw a bumblebee (*Bombus pennsylvanicus*) exploring prior to leaving a small patch
of *Baptisia australis*, where it had foraged nectar.

From the large body of descriptive information about the near-
target Exploration Flight in Aculeate Hymenoptera the following
series of reliable generalizations and predictive rules can be derived.

1.   The Exploration Flight always take place immediately after take-off and is followed either by straight-line departure or the return to the starting location.

2.   The Exploration Flight is invariably associated with targets to which an individual is likely to return, such as a nesting site or a resource that is not yet exhausted. No bee of any species has ever been seen to perform an Exploration Flight while flying from flower to flower for foraging. Yet bumblebees, as mentioned, have been seen to explore prior to leaving a patch of flowers.

3.   The Exploration Flight of all species comprises two highly distinct flight patterns, one for focal exploration and one for peripheral exploration.

4.   Focal exploration immediately follows takeoff. The explorer instantly turns around and faces the site of departure in about the same direction it will assume when returning and approaching at some later time (Collett & Baron, 1994). Then the explorer flies an oscillatory pattern. In most species these oscillations are arcuate left-right shifts of increasing amplitude, combined with gradually backing away from the target. The target or some nearby object lies near the center of these arcs.

If the target lies on a horizontal surface, the lateral oscillations expand into more or less complete circles around the target. If the target is stretched out vertically, like a stem or an artificial black stripe, it may be followed by irregular vertical movements up and down that can be accompanied by the lateral oscillations mentioned. In front of several nests of the stingless bee *Trigona iridipennis* (Bangalore, India), I observed, instead of the lateral oscillations as found in many other species of bees, mainly alternating forward-backward flight oscillations. During these forward-backward movements, these bees faced the nest entrance as other species of bees did during lateral oscillations. Whereas all observed and described focal search flights involved continuous displacement, the orchid bee *Euglossa cyanipes*, which I observed in Panama, acted differently. Females of this species, while exploring their nest location focally, alternate between stationary hovering flights and abrupt displacements laterally or vertically up or down.

5.   During peripheral exploration there are no oscillations and the explorers steadily fly forward, continuously curving alternately to

the left or right, passing through partial to full circles. The diameters of these circles gradually increase up to several meters and so does the elevation above the ground. If there is sufficient space and no wind, peripheral exploration symmetrically surrounds the target and facing covers all directions away from the target.

6.   A complete Exploration Flight always starts with focal exploration that then changes quickly but smoothly into peripheral exploration. The rhythm of oscillation during focal exploration continues without noticeable break into the rhythm of turn alternation during peripheral exploration. Peripheral exploration may be followed by a straight-flight departure, a return to the target or, rarely, by another cycle or cycles of focal and peripheral exploration. Sometimes either focal or peripheral exploration are performed in isolation.

All these rules are based on unaided observations. No complete Exploration Flight has ever been recorded and quantitatively analyzed. Species differ in various details of their Exploration Flight, but much of this interspecific differentiation has not yet been investigated.

Adaptive Function of the Aculeate Exploration Flight

All Aculeate Hymenoptera known to perform the Exploration Flight are also known to repeatedly visit a variety of locations within their home ranges. The targets at such locations may be one or several nesting sites, foraging sites, water holes or sites for collecting building material. The targets themselves, for the most part, are too small or hidden from direct sight to allow direct orientation (telotaxis) by means of direct perception by the approaching insect. Instead, targets are usually approached by indirect orientation, also called navigation. Such near-target topographic navigation, as found in a number of vertebrates and insects, makes use of memory for the spatial relations among landmarks around the targets. For Hymenoptera, the targets themselves, like the opening into a nest, may be minute in comparison to intergoal distances, which can span a number of kilometers in larger wasps and bees. For this, one would expect the use of different navigation mechanisms for proximal(terminal) navigation, the navigation while approaching a close target, and distal navigation for covering large stretches of space (Chmurzynki, 1964; Collett, 1992).

Importantly, the characteristic Aculeate Exploration Flight is performed exclusively near targets that are to be visited repetitively, like nesting sites and sustained sources of food or other resources. Bates' (1863) original suggestion that the Exploration Flight is a procedure for learning near-target (proximal) orientation cues (landmarks) has been convincingly confirmed by experiments with sphecid wasps and bees in a number of mutually corroborating studies (e.g. Baerends, 1941; Beusekom, 1948; Chmurzynski, 1964, 1967; Iersel, 1975; Iersel & Assem, 1964; Jander, 1994; Lehrer, 1991, 1993; Opfinger, 1931; Tinbergen, 1932; Tinbergen & Kruyt, 1938; Zeil, 1993a, b).

A detailed analysis of the near-target landmark knowledge and its acquisition by a sphecid wasp and the honeybees reveals an adaptive substructure: The closer the distance to the target, the finer the resolution of the remembered constellation of landmarks (Cheng, Collet, Pickhard & Wehner, 1987; Zeil, 1993b).

The decreasing resolution of landmark memory with distance from the nest entrance or from other targets is plausibly correlated with the structure of the Exploration Flight whose oscillations and loops gradually increase in amplitude as the distance from the target is increasing, thus covering a growing expanse of near-target space per unit time.

The Concepts "Behavioral Homology" and "Fixed Action Pattern"

The distinctive Exploration Flight of the Aculeate Hymenoptera is a new challenge for phylogenetic ethology: Here we have the first case of an unambiguously identifiable learning behavior whose phylogenetic distribution and evolutionary origin can be elucidated by analyzing homologies. Undeniably, all learning mechanisms, as all behavioral mechanisms, must have some genetic basis (Schneirla, 1956, p.133; Lorenz, 1965), and numerous genetic studies prove this correct (e.g., Wahlsten, 1978; Vargo & Hirsch, 1985; Brandes & Menzel, 1990; Tully, 1991). Shared genetic bases among learning mechanisms should establish homologies. However, discovering convincing homologies among learning mechanisms is impeded because they so often share analogous (homoplastic) similarities. The most notable examples for such homoplasies are the many striking similarities in the conditioning (causal learning) behavior Bitterman and his collaborators uncovered when they compared the learning of honeybees and vertebrates (see Bitterman & Couvillon, 1991).

Moreover, recognizing genetically specified constituents among behavioral traits and separating homologies from analogies is more of a challenge than it is in morphology because of the immense role nongenetic factors play in specifying behavior, as Schneirla (1956) had pointed out so forcefully. Early, keenly observing but not yet objectively quantifying ethologists discovered what are now known as "fixed action patterns," behavioral traits that they could convincingly homologize across taxa and that they considered as purely genetically (instinctively) specified traits (Heinroth, 1911; Lorenz, 1932; Whitman, 1898). The discovery of homology was correct; the perfect genetic attribution an overstatement, criticized with full justification by Schneirla (1956). No doubt, due to the uncanny capacity of our perceptual-cognitive system to automatically abstract rules from repeated observations, we tend to unconsciously ignore variability. After exact quantification had disclosed that so-called fixed action patterns do vary significantly and are therefore not truly fixed (Barlow, 1977; Schleidt, 1974), ethologists, by consensus, created a logical monster. Rather than defining the term "fixed action pattern" according to its original unrecognized abstract use, it was decided, and maintained to this day, that "fixed action pattern" means a "relatively fixed action pattern" (Hinde, 1982; McFarland, 1982; Schleidt, 1974; Thorpe, 1951). Barlow (1968, 1977) corrected the logical-terminological inconsistency by relabeling, turning the "fixed action pattern" into the "modal action pattern." This solved one logical dilemma but created another one: modal action patterns cannot be homologized because, by definition, they are partially specified by nongenetic information such as random information and plasticity, which is based on current sensory information and learned information. Only by a redefinition can we escape this conundrum.

Schneirla (1956), while criticizing the early European ethologists for dichotomizing observed behaviors into innate (instinctive) and acquired ones, was probably the first to recognize the above conundrum and to offer a logically clean solution (Schneirla, 1956, p. 138). Referring to detailed descriptions of feeding behavior in hydroid cnidarians, he recognized that this behavior is essentially stereotyped despite marked variability. Schneirla thus replaced the false phenomenological dichotomy of the early ethologists between fixed and acquired action patterns with the valid logical dichotomy between a species-specific stereotyped (fixed) essence and a plastic behavioral component of a behavior. I here adopt this very logical strategy and explicitly define the fixed action pattern as an abstract pattern, composed of the traits shared by multiple expressions of an identified action pattern, provided such sharing is not due to shared individual experience. Such sharing can be abstracted from multiple expressions of identified behaviors that are recorded from single individuals or from different individuals belonging to particular species or to higher taxa. The higher the taxon, the higher the expected degree of abstraction. I consider this definition of the fixed action pattern as logically identical with the original implicit-intuitive of Lorenz and with the idea of the stereotyped essence of Schneirla (1956). Should there be some danger of confusion between the recreated original definition and the current conventional definition of the fixed action pattern, the label "fixed action constituents" or "fixed action parameters" may substitute the label "fixed action pattern." Barlow's "modal action patterns," as observed and quantified in individuals and species, are then behavioral composites of the abstract fixed action patterns merged with nongenetically specified constituents that cause random and plastic variation.

True fixed action patterns or species typical behaviors develop instinctively (Schneirla, 1956, p. 177), hence are innately specified and can be homologized. In order to infer homologies among behavioral traits via noncircular logic, homology criteria have to be applied rigorously and systematically (Remane, 1956; Wiley, 1981). The fact that the Exploration Flight serves mainly, if not exclusively, the learning of local navigation cues is not a sufficient reason for rejecting a phylogenetic analysis based on finding homologies. This insight is fully consistent with Wenzel's (1992, p. 376) forceful statement: "There is no reason to discard any trait as useless to the illumination of homology and phylogeny because learning plays a role."

Still the best noncomputational and nonstatistical instructions for critically using homology criteria for the purpose of inferring phylogenies are found in Remane (1956). Baerends (1958), Wickler (1961), Lorenz (1981) and recently Wenzel (1992) explained and validated the application of Remane's rigorous methods to behaviors, especially the fixed action patterns. In the following phylogenetic analysis I employ criteria originally formulated by Remane (1956).

Macroevolution of the Aculeate Exploration Flight

All instances of the Exploration Flight in the Vespidae (paper wasps and relatives), the Spheciformes (digger wasps) and Apiformes (bees)—as taxonomically defined in Brothers (1975) and in Brothers and Carpenter (1993)—are mutually homologous for the following three major and logically independent reasons.

1.   Criterion of context: The Exploration Flight invariably occurs in the same behavioral context, that is, when leaving a location to which the individual habitually returns. In addition, there is experimental and observational evidence that the Exploration Flight is always incited either by the exposure to novel orientation cues or the failure to relocate a particular location as expressed by search behavior.

2.   Criterion of configuration or infrastructure: All instances of the Exploration Flight in all species in which it occurs share the same basic structure. There are two modular subunits—focal exploration and peripheral exploration—that can be employed separately or in combination, with focal exploration always preceding peripheral exploration. In addition, each of the two modular subunits has its own distinctive features, as described above. This high complexity adds weight to this argument: two unlearned behavioral processes are the more likely to be homologous the greater their complexity is.

3.   Criterion of taxonomic (cladistic) distribution: The Exploration Flight is a behavior that is restricted to the truly monophyletic suborder of the Aculeate (sting bearing) Hymenoptera. They have been observed in numerous flying nest-builders, and they are obviously absent in the flightless workers of ants. Among the millions of insect species outside the Aculeate Hymenoptera, not a single instance of the Exploration Flight is known. This fact permits the perfect application of the argument that instances of a morphological or behavioral character within a well-established taxon are the more likely to be homologous the less frequently this character is found

outside the taxon under discussion and the more frequently the character is present among the species inside this taxon. The application of this argument here is not circular because the taxon Aculeate Hymenoptera is defined on the basis of morphological characters that are not in any way tied to the character of the Exploration Flight (Brothers & Carpenter, 1993).

Given the fact that the Exploration Flight is restricted to Aculeate Hymenoptera, the question arises, where, within this taxon, is the origin of this behavior to be located? For this we have to combine the analysis of character homology with the cladistic analysis. The latter has clearly established three major lineages within the Aculeate Hymenoptera. First branched off the lineage of the Chrysidoidea, and thereafter the Vespoidea and Apoidea dichotomized (Brothers & Carpenter, 1993). The homologous character of the Exploration Flight bridges the two latter lineages, based on the evidence mentioned above. Hence, the Exploration Flight originated not later than the common root of the Vespoidea and Apoidea. This conclusion turns the attention to the earliest lineage of the Aculeate Hymenoptera, the Chrysidoidea. Species within this group do not build nests, but topographic orientation skills occur at least in the family of the Chrysididae (Rau, 1931; Rosenheim, 1987, 1989; Trexler, 1984). However, the question whether the Chrysididae possess the Exploration Flight is still largely open, except for the intriguing observation of stereotyped circling flights by *Argochrysis armilla* upon leaving the borrow of its *Ammophila* host (Rosenheim, 1987). This suggests the testable hypothesis that the Chrysidoidea may already possess the Exploration Flight of the Vespoidea and Apoidea. If this turns out to be true, then the Exploration Flight evolved near the phylogenetic root of the Aculeate Hymenoptera.

Elucidating cross-phyletic homologies is not the only way of shedding light on the phylogenetic origins of behavioral traits. Just as one can infer the origin of morphological structures by serial homology (a classic example is the appendages of a crayfish), one can similarly find in one species both an ancestral behavior and its evolutionary derivative. The prospects must be excellent for the Exploration Flight because it is phylogenetically not older than the Aculeate Hymenoptera.

Indeed, the search behavior of the Aculeate Hymenoptera must be homologous to their Exploration Flight for the following striking similarities, which are based on hundreds of observations by myself on dozens of species of Vespidae, Sphecifomes and Apiformes (bees). These observations conform with widely scattered published data that will not be explicitly cited.

1.   Aerial search flights, exactly as the Exploration Flight, have two highly distinct components or modules, focal and peripheral search. Both components of search flights are patterned in a way that clearly resemble their exploratory counterparts: During focal search as during focal exploration, the hovering wasps or bees focus on some distinct object and if they shift their position in space they do so predominantly by flying sideways. During peripheral search as during peripheral exploration, wasps and bees fly in forward direction, alternate between left- and right turns and cover much more ground than during focal search or focal exploration.

2.   Particular objects focused on during exploration are also focused on during search, and areas covered during exploration are also often covered during search. The converse, however, is often not the case. For instance, in foraging wasps and bees, focal search is often directed toward objects that are not the objects of exploration, and similarly, they often search over areas with alternating loops without covering them by peripheral exploration.

3.   Altering navigation cues near a permanent resource incites both searching and exploring, with the first preceding the second, as already mentioned above.

4.   For the honeybee it has been established that both focal search (Jacobs-Jessen, 1959; Menzel & Greggers, 1985) and focal exploration (Vollbehr, 1975) are accompanied by a tendency to face away from a source of light (negative phototaxis). In general, honeybee directional orientation during focal search is similar to their orientation during approach (Collett & Baron, 1994). A negative phototaxis during focal search has also been demonstrated in the andrenid bee *Panurgus banksianus* (Münster-Swendsen, 1968).

5.   Besides search, there is no other behavior known for wasps and bees that resembles in any way the Exploration Flight.

Given these five arguments, there is no doubt left that in the Aculeate Hymenoptera the aerial search and the Exploration Flight are homologous behaviors. This, then, raises the question, which one of the two behaviors is more ancestral? The answer is unambiguous; it must be search. Search itself is a behavioral universal, and there should be no doubt that all flying insects also search in flight, but only among the Aculeate Hymenoptera do we find both search flights and the Exploration Flight. Also, within the Aculeate Hymenoptera search flights are employed more frequently and in more contexts than the Exploration Flight. As mentioned above, some search flights are seen at locations where the Exploration Flight is not seen, but all areas covered by the Exploration Flight are sometimes covered by search flights.

Despite all these similarities, search flights and the Exploration Flight are so distinct that the experienced observer can tell them easily apart after a few seconds of observation: Search flights are noticeably less stereotyped than the Exploration Flight. During search flights, bees and wasps tend to approach objects; during the Exploration Flight they tend to recede from them. Finding the nest or a resource terminates searching; most instances of the Exploration Flight are terminated by departure.

In most bees and wasps alternating left- and right-turns shape similar figure-eight-like patterns during search as during peripheral exploration. A marked deviation from this similarity occurs within the monophyletic lineage of the corbiculate bees that comprises the orchid bees (Euglossini), the stingless bees (Meliponini), the bumblebees (Bombini) and the honeybees (Apini) (Roig-Alsina & Michener, 1993).

In all the stingless bees (*Trigona iridipennis*, *T. angustula*, *Paratrigona opeca*, *Melipona fasciata)* and one species of orchid bee *(Euglossa cyanipes)* that I observed, the peripheral exploration flights still comprised the prototypical figure-eight-like elements. In the bumblebees (*Bombus pennsylvanicus*, *B.lapidarius*, *B.terrestris*) and in the honeybees (*Apis mellifera*, *A. cerana*, *A. dorsata*, *A. florea*), the peripheral exploration pattern deviates markedly from the peripheral search pattern. Their peripheral exploration flights are performed in the form of spirals that expand and increase in height. Within these spirals the bees alternate between clockwise and counterclock-

wise smooth turns or arcs. Alternating turning arcs are connected by
U-turns. In other words, there is a conspicuous alternation between
sharp and smooth turns. The peripheral search of honeybees and
bumblebees, like that of other bees, is composed of figure-eight-like
elements with alternating left-right turns that have the same shape
and that thus differ from the alternating sharp and smooth turns of
their peripheral exploration flight.

Flowers, Landmarks and Categories of Learning

The Exploration Flight of the Aculeate Hymenoptera must be considered an instance of pure learning behavior because its only known
function is to systematically collect and store in long- term memory
information about the spatial constellation of local landmarks. This
stored information is used to implement, after a long-distance flight,
the terminal navigation procedures for a pinpoint landing on a target. This set of features, then, raises the question, where in the traditional explanatory and conceptual framework of learning behaviors
does the Exploration Flight fit?

The answer is confusingly ambiguous, depending on the criteria one pays attention to. Learning navigation cues by exploration is
often referred to as "latent learning," a "nonassociative" form of learning. Latent learning does not require an explicit reinforcer, and, indeed, unrewarded bees and wasps that leave their nests for the first
time perform the Exploration Flight. For these very reasons, learning the local landmarks might be considered an instance of
nonassociative latent learning, as has been pointed out by Menzel
(1993). However, at artificial feeding stations, departing honeybees
explore locally and learn local landmarks after a major food reward.
Hence, this learning context fits the criterion of "backward conditioning," as has been pointed out by Lehrer (1993). Backward conditioning, however, is a subcategory of associative learning.

Since bees that leave food somehow associate the reward with
surrounding visual cues, one could speak of Pavlovian conditioning,
and because they also learn the visually controlled approach maneuvers, one could speak of instrumental (operant) conditioning
(Catania, 1984; Mackintosh, 1974). Finally, since the Exploration
Flight is decreasingly expressed in the course of repetitive exposure
to the same local cues, there is an argument to speak of habituation

(Peeke & Herz, 1973). Finally, in frustration, one could dismiss all of the above because of this behavior's complexity and speak of cognition or cognitive learning (Gallistel, 1989), which is not satisfactory either, because there is no learning that could be truly called noncognitive. To avoid this conceptual dilemma, let me propose two others schemes of classification:

First, one can distinguish between learning universals or widely shared learning processes—as focused on by Bitterman (1988) and Bitterman and Couvillon (1991)—and specific or "customized" learning—as focused on by Gould (1984). In the light of this conceptual dichotomy, the Aculeate Exploration Flight is clearly "customized" by virtue of its absolute restriction to a monophyletic taxon and by virtue of its function to specifically learned cues to be used for proximal navigation.

Second, one can use the honeybee's learning dichotomy for contextual (landmark) learning and item (rewarding object) learning, as was proposed by Lehrer (1993). While approaching or leaving a feeding station, an individual bee learns the spatial context of nearby landmarks (Collett & Baron, 1994; Couvillon & Bitterman, 1992; Lehrer, 1993; Opfinger, 1931). From this contextual learning we can conceptually and operationally separate item learning, specifically the learning of a reinforcing object. This associative learning process has critical characteristics of Pavlovian conditioning: spatiotemporal proximity between US and CS is important, and backward conditioning does not work, as is observed when honeybees learn the odor or color of an artificial feeder (Opfinger, 1931, 1949; Menzel, 1983; Menzel & Bitterman, 1983). The precise separation or area of overlap between item learning and contextual learning has not yet been experimentally analyzed (Lehrer, 1993).

Summary

Exploration is the learning of topographic knowledge to be used for finding ways within home ranges. Topographic curiosity is the underlying drive whose incentive is topographic novelty. Exploratory behavior evolved at least three times independently in the mollusks, in the arthropods and in the vertebrates. Within the Aculeate (stinging) Hymenoptera, the distinct behavior of the Exploration Flight evolved. The Exploration Flight is performed when leaving a target site worth returning to. Difficulty of finding a target site, or the experience of novel landmarks surrounding it, incite the Exploration Flight. Near-target landmarks are learned by it. The Aculeate Exploration Flight is composed of two behavioral modules that can be employed separately or in combination: During focal exploration the bee or wasp focuses continually on the target or some nearby object while performing oscillatory flight maneuvers; during peripheral exploration the bee or wasp flies ahead in gradually expanding loops around the target. The Exploration Flight meets the criteria of the ethological concept of the "fixed action pattern" as well as Schneirla's near-identical concept of the "stereotyped essence." It is possible to reconstruct the phylogeny of the Exploration Flight by applying the homology criteria of Remane. The Exploration Flight evolved once early in the evolution of the Aculeate (stinging) Hymenoptera out of their aerial search behavior. The Exploration Flight falls into a category of learning that serves the acquisition of spatiotemporal contextual knowledge, which is distinct from the category of conditioning-like learning of the visual, chemical and perhaps tactile properties of targeted objects themselves.

Acknowledgments

For hospitality and help in the study of exploratory behavior in various Hymenoptera, I thank F. Dyer, R. Gadagkar, N. Koeniger and D. Roubik. For advice on taxonomic and phylogenetic issues, I thank B. Alexander and C.D. Michener. For challenging my thinking about learning behavior, I thank M.E. Bitterman. For guidance towards some publications of Schneirla, I thank G. Greenberg. A portion of my research on exploration flights was supported by the General Research Fund of the University of Kansas.

References

Akre, R. D., Garnett, W. B., MacDonalds, J. F., Greene, A., & Landolt, P. (1976). Behavior and colony development of *Vespula pennsylvanica* and *V. atropilosa* (Hymenoptera: Vespidae). *Journal of the Kansas Entomological Society, 49,* 63–84.

Arbas, E. A., Meinertzhagen, I. A., & Shaw, S. R. (1991). Evolution in nervous systems. *Annual Review of Neuroscience, 14,* 9–38.

Archer, J. & Birke, L. I. A. (eds.). (1983). *Exploration in animals and humans.* Wokingham, Berkshire, England: Nostrand Reinhold.

Baerends, G. P. (1941). Fortpflanzungsverhalten und Orientierung der Grabwespe *Ammophila campestris. Tijdschrift Entomologie, 84,* 68–275.

Baerends, G. P. (1958). Comparative methods and the concept of homology in the study of behaviour. *Archives Néerlandaise De Zoologie, 13,* 401–417.

Barlow, G. W. (1968). Ethological units of behavior. In D. Ingle (ed.), *The central nervous system and fish behavior* (pp. 217–232). Chicago: University of Chicago Press.

Barlow, G. W. (1977). Modal action pattern. In T. A. Sebeok (ed.), *How animals com-*

municate (pp. 98–134). Bloomington, Ind.: Indiana University Press.

Barnett, S. A. (1975). *The rat: A study in behavior.* Chicago: University of Chicago Press.

Bates, H. W. (1863). *The naturalist on the River Amazon* Vol. 2. London: J. Murray.

Becker, L. (1958). Untersuchungen über das Heimfindevermögen der Bienen. *Zeitschrift für vergleichende Physiologie, 41,* 1–15.

Berlyne, D. E. (1950). Novelty and curiosity as determinants of exploratory behaviour. *British Journal of Psychology, 41,* 68–80.

Berlyne, D. E. (1960). *Conflict, arousal and curiosity.* New York: McGraw-Hill.

Berlyne, D. E. (1966). Curiosity and exploration. *Science, 153,* 25–33.

Bethe, A. (1898). Dürfen wir den Bienen und Ameisen psychische Qualitäten zuschreiben? *Pflügers Archiv für die gesamte Physiologie, 70,* 15–100.

Beusekom, G. v. (1948). Some experiments on the optical orientation in *Philanthus triangulum* Fabr. *Behaviour, 1,* 3–4 (195–225).

Bitterman, M. E. (1988). Vertebrate-invertebrate comparisons. In H. J. Jerison & I. Jerison (eds.), *Intelligence and evolutionary biology* (pp. 251–275). New York: Springer.

Bitterman, M. E., & Couvillon, P. A. (1991). Failures to find evidence of adaptive specialization in the learning of honeybees. In L. J. Goodman & R. C. Fisher (eds.), *The behaviour and physiology of bees* (pp. 302–322). Wallingford, UK: CAB International.

Blodgett, H. C. (1929). The effect of introduction of reward upon the maze performance of the rat. *University of California Publications in Psychology, 4,* 113–134.

Bovet, J. (1992). Mammals. In F. Papi (ed.), *Animal Homing* (pp. 321–361). New York: Chapman & Hall.

Bowring, S. A., Grotzinger, J. P., Isachsen, C. E., Knoll, A. H., Pelechaty, S. M., & Kolosov, P. (1993). Calibration rates of early cambrian evolution. *Science, 261,* 1293–1298.

Brandes, C., & Menzel, R. (1990). Common mechanisms in proboscis extension conditioning and visual learning revealed by genetic selection in honeybees (*Apis mellifera*). *Journal of Comparative Physiology A, 166,* 545–552.

Brothers, D. J. (1975). Phylogeny and classification of the aculeate Hymenoptera, with special reference to Mutillidae. *University of Kansas Science Bulletin, 50,* 483–648.

Brothers, D. J., & Carpenter, J. M. (1993). Phylogeny of the aculeata: Chrysidoidea and Vespoides (Hymenoptera). *Journal of Hymenoptera Research, 2,* 227–304.

Buttel-Reepen, H. v. (1915). *Leben und Wesen der Biene.* Braunschweig: Vieweg.

Catania, A. C. (1984). *Learning* 2nd ed. Engelwood Cliffs, N.J.: Prentice-Hall.

Cheng, K., Collet, T. S., Pickhard, A., & Wehner, R. (1987). The use of visual landmarks by honey bees: bees weight landmarks according to their distance from the goal. *Journal of Comparative Physiology, A 161,* 469–475.

Chmurzynski, J. A. (1964). Studies on the stages of spatial orientation in female *Bembex rostrata* (Linné 1748) returning to their nests (Hymenoptera, Sphecidae). *Acta Biologica Experimentalis (Warsaw), 24,* 103–132.

Chmurzynski, J. A. (1967). On the role of relations between landmarks and the nesthole in the proximate orientation of female *Bembex rostrata* (Linné) (Hymenoptera, Sphecidae). *Acta Biologica Experimentalis (Warsaw), 27,* 221–254.

Collett, T. S. (1992). Landmark learning and guidance in insects. *Philosophical Transactions of the Royal Society of London. Ser. B, 337,* 295–303.

Collett, T. S., & Baron, J. (1994). Biological compass and the coordinate frame of landmark memories in honeybees. *Nature, 368,* 137–140.

Couvillon, P. A., & Bitterman, M. E. (1992). Landmark learning by honeybees. *Journal of Insect Behavior, 6,* 123–129.

Dickinson, A., & Balleine, B. (1994). Motivational control of goal-directed action. *Animal Learning and Behavior, 22,* 1–18.

Evans, H. (1966). *The comparative ethology and evolution of the sand wasps.* Cambridge, Mass.: Harvard University Press.

Fowler, H. (1965). *Curiosity and Exploratory Behavior.* New York: Macmillan.

Freisling, J. (1943). Zur Psychologie der Feldwespe. *Zeitschrift für Tierpsychologie, 5,* 438–463.

Frisch, K. von (1967). *The dance language and orientation of bees.* Cambridge, Mass.: Belknap Press of Harvard University Press.

Frison, T. H. (1930). Observations on the behavior of bumble bees (*Bombus*). The orientation flight. *Canadian Entomologist, 62,* 49–54.

Gallistel, C. R. (1989). *The organization of learning.* Cambridge, Mass.: Bradford Books/MIT Press.

Gibson, E. J. (1988). Exploratory behavior in the development of perceiving, acting, and the acquiring of knowledge. *Annual Review of Psychology, 39,* 1–41.

Gould, J. L. (1984). Natural history of honey bee learning. In P. Marler & H. S. Terrace (eds.), *The biology of learning* (pp. 149–180). New York: Springer.

Halliday, M. S. (1968). Exploratory behaviour. In L. Weiskrantz (ed.), *Analysis of behavioural change* (pp. 106–126). New York: Harper & Row.

Heinroth, O. (1911). Beiträge zur Biologie, namentlich ethologie und Psychologie der Anatiden. In *Verhandlungen des V. Internationalen Ornithilogen-Kongresses* (pp. 589–702).

Hinde, R. A. (1982). *Ethology. Its nature and relations with other sciences.* New York: Oxford University Press.

Honig, W. K. (1987). Local cues and distal arrays in the control of spatial behavior. In P. Ellen & C. Thinus-Blanc (eds.), *Cognitive processes and spatial orientation in animals and man* (pp. 73–88). Netherlands: Martinus Nijhoff.

Hudson, B. B. (1950). One-trial learning in the domestic rat. *Genetic Psychology Monographs, 41,* 99–145.

Iersel, J. J. A. van. (1975). The extension of the orientation system of *Bembex rostrata* as used in the vicinity of its nest. In G. Baerends, C. Beer & A. Manning (eds.), *Function and evolution of behaviour* (pp. 142–168). Oxford: Clarendon Press.

Iersel, J. J. A., van, & Assem, J., van den. (1964). Aspects of orientation on the digger wasp *Bembix rostrata. Animal Behaviour, 1 Suppl.,* 145–162.

Jacobs-Jessen, U. F. (1959). Zur Orientierung der Hummel und einiger anderer Hymenopteren. *Zeitschrift für vergleichende Physiologie, 41,* 597–641.

Jander, R. (1975). Ecological aspects of animal orientation. *Annual Review of Ecology and Systematics, 6,* 171–188.

Jander, R. (1990). Exploration and search in foraging honey bees (Apis mellifera) and paper wasps (Vespula maculifrons). In G. K. Veeresh, B. Mallik, & C. A. Viraktamath (eds.), *Social insects and the environment. Proceedings of the 11th international congress of IUSSI* (pp. 570–571). New Delhi: Oxford & IBH.

Jander, R. (1994). Observations and experiments on exploration in the Hymenoptera, unpublished.

Janzen, D. H. (1971). Euglossine bees as long-distance pollinators of plants. *Science, 171,* 203–205.

Keith-Lucas, T., & Guttman, N. (1975). Robust-single-trial backward conditioning. *Journal of Comparative Physiological Psychology, 88,* 468–476.

Kemper, H., & Döhring, E. (1967). *Die sozialen Faltenwespen Mitteleuropas.* Berlin: Parey.

Klinksik, C., & Schönitzer, K. (1988). Lebensweise und Verhalten der solitären Sandbiene *Andrena nycthemera* (Hymenoptera, Andrenidae). *Verhandlungen der Deutschen Zoologischen Gesellschaft, 88,* 84.

Lehrer, M. (1991). Bees which turn back and look. *Naturwissenschaften, 78,* 274–276.

Lehrer, M. (1993). Why do bees turn back and look? *Journal Comparative Physiology A, 172,* 549–563.

Lorenz, K. (1932). Betrachtungen über das Erkennen der arteigenen Triebhandlungen bei Vögeln. *Journal für Ornithologie, 80,* 50–98.

Lorenz, K. Z. (1965). *Evolution and modification of behavior.* Chicago & London: University of Chicago Press.

Lorenz, K. Z. (1981). *The foundations of ethology.* New York: Springer-Verlag.

Mackintosh, N. J. (1974). *The psychology of animal learning.* New York: Academic Press.

Manning, A. (1956). Some aspects of foraging behaviour of bumblebees. *Behaviour, 9,* 164–202.

McFarland, D. (1982). *The oxford companion to animal behavior.* New York: Oxford University Press.

Menzel, R. (1983). Neurobiology of learning and memory: The honeybee as a model system. *Naturwissenschaften, 70,* 504–511.

Menzel, R. (1993). Associative learning in honey bees. *Apidology, 24,* 157–168.

Menzel, R., & Bitterman, M. E. (1983). Learning by honeybees in an unnatural situation. In F. Huber & H. Markl (eds.), *Neuroethology and Behavioral Physiology* (pp. 206–215). New York: Springer.

Menzel, R., & Greggers, U. (1985). Natural phototaxis and its relationship to colour vision in honey bees. *Journal of Comparative Physiology, 157,* 311–321.

Michener, C. D. (1953). The biology of a leafcutter bee, *Megachile brevis* and its associates. *University of Kansas Science Bulletin, 35,* 1659–1748.

Michener, C. D. (1966). The bionomics of a primitively social bee, *Lasioglossum versatum,* (Hymenoptera: Halictidae). *Journal of the Kansas Entomological Society, 39,* 193–217.

Miklos, G. L. G. (1993). Molecules and cognition: the latterday lessons of levels, language, and lac. Evolutionary overview of brain structure and function in some vertebrates and invertebrates. *Journal of Neurobiology, 24,* 842–890.

Morris, S. C. (1993). The fossil record and the early evolution of the metazoa. *Science, 316,* 219–225.

Münster-Swendsen, M. (1968). On the biology of the solitary bee *Panurgus banksianus* Kirby (Hymenoptera, Apidae) including some ecological aspects. In *Royal Vetererinary and Agricultural College Copenhagen, Denmark Yearbook* Vol. 1968 (pp. 215–241).

Opfinger, E. (1931). Über die Orientierung der Biene an der Futterquelle (Bedeutung von Anflug und Orientierungsflug für den Lernvorgang bei Farb- Form- und Duftdressur.). *Zeitschrift für vergleichende Physiologie, 15,* 431–487.

Opfinger, E. (1949). Zur Psychologie der Duftdressuren bei Bienen. *Zeitschrift für vergleichende Physiologie, 31,* 441–453.

Papi, F. (ed.). (1992). *Animal homing.* New York: Chapman & Hall.

Pearce, J. M. (1987). *Introduction to animal cognition.* Hillsdale NJ: Lawrence Erlbaum Associates.

Peeke, H. V., & Herz, M. J. (eds.). (1973). *Habituation* Volume 1: Behavioral studies. New York: Academic Press.

Plath, O. E. (1934). *Bumblebees and their ways.* New York: Macmillan.

Poucet, B. (1993). Spatial cognitive maps in animals: New hypotheses on their structure and neural mechanisms. *Psychological Review, 100,* 163–182.

Rajotte, E. G. (1979). Nesting, foraging and pheromone response of the bee *Colletes validus* Cresson and its associations with low brush blueberries (Hymenoptera: Colletidae) (Ericaceae: Vaccinium). *Journal of the Kansas Entomological Society, 52,* 349–361.

Rau, P. (1931). Notes on the homing of several species of wasps (Hyn.: Chysididae, Sphegoidae, Vespoidae). *Entomological News, 42,* 199–200.

Remane, A. (1956). *Die Grundlagen des natürlichen Systems der vergleichenden Anatomie and Phylogenetik* 2nd ed. Leipzig: Geest und Portig K.G.

Roig-Alsina, A., & Michener, C. (1993). Studies of the phylogeny and classification of long-tongued bees (Hymenoptera: Apoidea). *University of Kansas Science Bulletin, 55,* 123–173.

Rosenheim, J. A. (1987). Host location and exploitation by the cleptoparasitic wasp

Argochrysis armilla (Hymenoptera: Chrysididae): The role of learning. *Behavioral Ecology and Sociobiology, 21,* 401–406.

Rosenheim, J. A. (1989). Behaviorally mediated spatial and temporal refuges from a cleptoparasitic, *Argochrysis armilla* (Hymenoptera: Chrysididae), attacking a ground-nesting wasp, *Ammophila dysmica* (Hymeoptera: Sphecidae). *Behavioral Ecology and Sociobiology, 25,* 335–348.

Russel, P. A. (1973). Relationship between exploratory behaviour and fear. *British Journal of Psychology, 64,* 417–433.

Schleidt, W. M. (1974). How "fixed" is the fixed action pattern? *Zeitschrift für Tierpsychologie, 36,* 184–211.

Schneirla, T. C. (1941). Studies on the nature of ant learning. I. The characteristics of a distinctive initial period of generalized learning. *Journal of Comparative Psychology, 32,* 41–82.

Schneirla, T. C. (1943). The nature of ant learning. II. The intermediate stage of segmental maze adjustment. *Journal of Comparative Psychology, 35,* 149–176.

Schneirla, T. C. (1956). Interrelationships of the "innate" and the "acquired" in instinctive behavior. In P. Grassé (ed.), *Instinct dans le comportement des animaux et de l'homme* (pp. 131– 187). Paris: Masson et Cie.

Schöne, H. (1984). *The spatial control of behavior in animals and man (translated from German).* Princeton, N.J.: Princeton University Press.

Seeley, T. D. (1985). *Honeybee ecology.* Princeton N.J.: Princeton University Press.

Thinus-Blanc, C., Save, E., Buhot, M.-C., & Poucet, B. (1991). The hippocampus, exploratory activity, and spatial memory. In J. Paillard (ed.), *Brain and space* (pp. 334–352). New York: Oxford University Press.

Thorpe, W. H. (1951). The definition of terms used in animal behaviour studies. *Bul. Animal Behavour, 9,* 34–40.

Tinbergen, N. (1932). Über die Orientierung des Bienenwolfes (*Philanthus triangulum* Fabr.) *Zeitschrift für vergleichende Physiologie, 16,* 305–334.

Tinbergen, N., & Kruyt, W. (1938). Über die Orientierung des Bienenwolfes (*Philanthus triangulum* Fabr.) III. Die Bevorzugung bestimmter Wegmarken. *Zeitschrift für vergleichende Physiologie, 25,* 292–334.

Trexler, J. C. (1984). Aggregation and homing in a chrsidid wasp. *Oikos, 43,* 133–137.

Tully, T. (1991) Physiology of mutations affecting learning and memory in *Drosophila*— the missing link between gene product and behavior. *Trends in Neurosciences. 14,* 163–164.

Vargo, M., & Hirsch, J. (1985) Selection for central excitation in *Drosophila melanogaster. Journal of Comparative Psychology, 99,* 81–86.

Vollbehr, J. V. (1975). Zur Orientierung junger Honigbiene bei ihrem ersten Orientierungsflug. *Zoologische Jahrbücher. Abteilung für Allgemeine Zoologie und Physiologie der Tiere, 79,* 33–69.

Voss, H. G., & Keller, H. (1983). *Curiosity and exploration. Theories and results.* New York: Academic Press.

Wahlsten, D. (1978). Behavioral genetics and animal learning. In H. Anisman & G. Bignami (eds.), *Psychopharmacology of aversively motivated behavior,* (pp. 63– 118). New York: Plenum Press.

Wcislo, W. T. (1992). Nest localization and recognition in a solitary bee, *Lasioglossum (Dialictus) figuerresi* Wcislo (Hymenoptera: Halictidae) in relation to sociality. *Ethology, 92,* 108– 123.

Wehner, R. (1992). Arthropods. In F. Papi (ed.), *Animal homing* (pp. 45–144). New York: Chapman & Hall.

Wenzel, J. W. (1992). Behavioral homology and phylogeny. *Annual Review of Ecology and Systematics, 23,* 361–381.

Whitman, C. O. (1898). Animal behavior. In *Biological lectures of the marine biological laboratory.* Woods Hole, Mass.: Marine Biological Laboratory.

Wickler, W. (1961). Ökologie und Stammesgeschichte von Verhaltensweisen. *Fortschritte der Zoologie, 13,* 303–365.

Wiley, E. O. (1981). *Phylogenetics*. New York: Wiley.

Wolf, E. (1926). Über das Heimkehrvermögen der Bienen I. *Zeitschrift für vergleichende Physiologie, 3*, 615–691.

Zeil, J. (1993a). Orientation flights of solitary wasps (*Cerceris*; Sphecidae; Hymenoptera) I. Description of flight. *Journal of Comparative Physiology A, 172*, 189–205.

Zeil, J. (1993b). Orientation flights of solitary wasps (*Cerceris*; Sphecidae; Hymenoptera. II. Similarities between orientation and return flights and the use of motion parallax. *Journal of Comparative Physiology A, 172*, 207–222.

Desert Ants
Their Evolution and Contemporary
Characteristics

G. M. Dlussky

Characteristics of the Desert

The animals and plants of deserts have long elicited the attention of researchers. In recent years, the number of desert studies has dramatically risen, chiefly because of the necessity to better manage arid lands. However, the biological literature is mainly statistical, encompassing productivity figures from various desert biogeocenoses, ways to increase their yields and the morphophysiological adaptations that enable the animals and plants to survive under such harsh conditions. The evolutionary approach, as a rule, is totally absent. And yet it is only this approach that makes it possible to understand why those and no other adaptations, or why any particular and no other biogeocenotic structures, can occur. The study of the ways various organisms become adapted to life in deserts also can give us extremely rich material for learning how to measure phylogenesis in general. It is no coincidence that aquatic organisms (sharks, ichthyosaurs, whales, skates and flounders) and animals from arid lands (hamsters, kangaroo rats, gophers, mole rats and sand rats) are classical examples of convergence.

As Kashkarov correctly remarked (Kashkarov & Korovin, 1936, p. 34), "in the desert climate, there are constant challenges to the organisms' satisfying their basic needs. The organisms are somehow obliged, by changing either their structure, or their behavior, or yet their physiological characteristics, to adapt themselves to the desert conditions: thus, natural selection in the desert often acts in the most cruel ways."

The basic general challenging factors, common to all the deserts on Earth, are the following:

1. Insufficient humidity.
2. Harsh temperature conditions.
3. Scarcity of food.

There are many definitions of the term "desert." I agree with the description by Petrov (1973, p. 8), according to whom, the term "desert" includes

territories with an extremely dry climate, i.e, precipitation less than 200 or 250 mm per year, which is irregular and with an evaporation that results in occasional slight humidity. Sometimes there is a network of dead riverbeds or ephemeral streams, but there is never any permanent flow on the soil's surface, with soil fissures or salty areas characterized by the production of sulphides and chlorides. In such territories, where the life of organisms is often suppressed, the plant cover and the animal world are very scarce and, in such soils, the migration of salts in solution is more important than the biogenetic processes in all areas where agriculture can be practiced only with irrigation.

Diverse Associations and Coadaptive Complexes

While studying the structure of any biocenosis, including desert ants, one can find in it groups of organisms related to each other by competitive and symbiotic "bonds." The most elementary units of such a structure are the *multispecies associations*. The term "multispecies association" was proposed by A. V. Demchenko (1975) and became widely adopted in Russian myrmecological literature (Zakharov, 1978, 1991; Reznikova, 1976).

Members of a multispecies association live together in a neighborly evolutionary fashion, occupying ecological niches, which often overlap. For this reason, a change in the quantity or activity of any single species causes a change in the quantity or activity of many other members of the multispecies association, and it consequently modifies the very structure of the whole association. Often, but not necessarily, the multispecies association is organized in a hierarchical fashion: the dominant species influences the quantity and activity of all subordinates. Examples of multispecies associations are: the herds of hoofed animals in the African savannas; the mixed flocks of birds wintering in Russian woods, such as herons, cranes and passerines; ants associated with dominant ant species, such as ants nesting within the territories of *Formica pratensis* and *Formica rufa*.

We propose to name any group of species inhabiting the same biocenosis and interconnected by competitive and mutualistic relations a *coadaptive complex* (Dlussky, 1975a, 1981a, b). The coadaptive complex, therefore, is the sum total of all closely related multispecies associations. For example, a single mixed flock of passerines and other bird species is a multispecies association. In a particular forest, all the flocks are a coadaptive complex. A broader classification is the *zonal complex*, or the sum of all related coadaptive complexes within a single natural biome; for example, the flocks of birds wintering in the taiga zone, or the desert ants of the Palearctic deserts.

The proposed classification system reflects various levels of looking at the same events: the individual level, the population level, and the species level. In a multispecies association, the functioning interacting elements are the individual elements, such as individual insects or colonies of social insects. In the coadaptive complex, the individual elements are the populations; and in the zonal complex, the species are the individual elements. When all the members of a certain taxon possess characteristics that enable them to escape competition with members of other taxa, such as hunting food at night, as in the case of bats (McNab, 1971), and sociality, as in ants and termites (Dlussky, 1975a; Davidson, 1976a), the coadaptive complex consists of only members of that taxon. But, in many cases, coadaptive complexes consist of members of different taxa, such as flocks of wintering birds and herds of African hoofed animals.

The structure of ant coadaptive complexes is very different in the various biomes. In the humid biocenoses of the Palearctic region, the food intake was essentially the same for all ant species: they consumed insect bodies; live, slow-moving soft-skinned bugs; excrements of *Homoptera* (mostly aphids); and various plant seeds. It is possible at present to identify many life forms (Arnoldi, 1937; Seyma, 1972), which seek their food on different levels of the ecosystem: for example, the geobionts from the soil; the stratobionts from the leaf litter; the herpetobionts from the soil's surface; the hortobionts from the grass; and the dendrobionts from the trees. The independent group consisting of the dominant species *Formica rufa*, *Lasius fuliginosus* and *Liometopim* spp get their food from several biogeocenotic levels. They are characterized by the great quantity of their individuals, i.e., up to several million; by their large food-gathering territory, i.e., up to several hectares, which is well protected against intrusions by extraterritorial ants; and by a fully developed road system. Colonies of these dominants serve as organizers for multispecies associations, and the structure of these species' territories determines how the other species' nests will be distributed (Kaczmarek, 1953; Stebaev, Yamkovoy, Stepanova, Milovidova & Tarasenko, 1967; Stebaev, 1971; Reznikova, 1976).

The combination of species in each particular biocenosis is determined primarily by their historical sequence and the peculiarities of the local environment, whereas the mutual interrelationship of the various life forms depends upon the growth characteristics of plants. Thus, in the woods around Moscow the multispecies association most often looks as follows: the dominant forms occupying several levels of complexes are *Formica rufa*, *F. aquilonia* or *Lasius fuliginosus*; the stratobiont *Myrmica rubra*; the herpetobionts *Lasius niger* and *Formica fusca*; the dendrobiont *Camponotus herculaneus*. In the oak forest of Crimea's mountains, the dominant species is *Formica gagates*; *Aphaenogaster subterranea* is a stratobiont; *Lasius emarginatus* is a herpetobiont; and both *Crematogaster scutellaris* and *Camponotus lateralis* are dendrobionts. In the humid woodland ecosystem, the most numerous forms are the herpetobionts and the stratobionts; other species are less numerous. In the grassland biocenoses, the geobionts are most often more frequent than the stratobionts and herpetobionts. Such a structure of the coadaptive complexes has been functioning in the Palearctic since at least the Miocene period (Dlussky, 1981a).

The structure of coadaptive complexes of ants from tropical forests is distinguished by its complexity. Judging by our observations on Tonga and on Western Samoa (Polynesia), the ant population in tropical woods is divided into two quite isolated groups: one of them builds nests in the soil and another on trees. Among the members of the first group, only a few species are real herpetobionts, e.g., in Tonga's native communities, only *Odontomachus simillimus* was found. The majority of the species are geoand stratobionts. They hunt in both the leaf litter and in the soil itself, as well as on its surface. This group is far more numerous and diversified in the deep inland of continental tropics than on islands. In the continental rain forests of Asia, Africa and South America, ants are many times more numerous per square meter than on any island.

More than half of the ant nests in tropical forests are found on trees. It is noteworthy that every tree in a tropical forest is festooned by lianas and adorned by a multitude of epiphytes, including plant species with the most diverse life modes. Among the epiphytes on the dead branches of trees are encountered the same species that inhabit the leaf litter: in Polynesia, e.g., *Plagiolepis alluaudi* and *Solenopsis papuana*, as well as some species similar to typical stratobionts, belonging to the genera *Ponera*, *Hypoponera* and *Strumigenys*. Many species, such as *Pseudomyrmicinae*, *Acanthostichus*, *Cylindromyrmex*, possess special adaptations enabling them to burrow into wood: a cylindrical body, plus modified jaws and head shape. The nest population on tropical trees, as we have observed, is always a multispecies one.

On the other hand, the structure of ant coadaptive complexes in arid regions is totally different from those in humid climates (Dlussky, 1975a, 1981b). The paucity of ant fauna in the soil and the lack of woody flora cause all desert ants to be herpetobionts. As a result, the biocenotic combination of species is achieved at the expense of food specialization.

There are three main trophic groups of ants in deserts: diurnal zoonecrophages or necrophages, nocturnal zoophages and carpophages (Dlussky, 1975a, 1981b). The first group has conserved the food habits of humid zone ants, i.e., these ants from dawn to dusk hunt living insects or collect dead insect bodies and consume the excrement of *Homoptera*. But, in deserts such habitats demand special adaptations as the soil surface is very hot during the day. Diurnal zoonecrophages must move very quickly to investigate their feeding territory and return to the nest as soon as possible. All large- and medium-sized species of this group have long legs and slender bodies, e.g., *Cataglyphis* and *Acantolepsis* in Palearctic deserts; *Myrmecocystus* and *Conomyrma* in North America; *Melophorus* in Australia. Often they also lift their gaster vertically during running. Small necrophages, e.g., *Plagiolepis* and *Brachymyrmex*, do not differ morphologically from the ants of the humid zone, but they usually collect food within the shadow of plant foliage.

The prey of nocturnal zoophages (*Camponotus* in Palearctic region) is quite different from that of the diurnal group, for at night many soft-bodied insects, e.g., termites, cockroaches and some larvae, appear on the soil surface. Night hunting demands unusual special adaptations for nocturnal orientation that are not typical in ants of the humid zone.

Carcophages (seed eaters) are represented in all the deserts by different genera of the subfamily *Myrmicinae*, e.g., *Messor* and *Monomorium* in Palearctic deserts; *Pogonomyrmex* and *Veromessor* in North America; *Solenopsis* in South America; *Chelaner* in Australia. *Pheidole* are found in all deserts. Most of these ants are polymorphous and the major workers (soldiers) have large heads with massive jaws. The characteristics of this structure are related to the crushing of hard seeds.

Each trophic group consists of a series of different sizes. Ants of every size collect prey, and the size of the prey is related to the size of the individual ant (Dlussky, 1975a; Davidson, 1976a, b). A fourth trophic group, fungus-growing ants (*Trachymyrmex, Mycetophylax*), are found only in American deserts.

Ant complexes in semiarid regions may be either herpetobionts or dendrobionts. In the European steppe communities, an example of a semiarid situation, the dominant species are all herpetobionts and the food composition favored by each of those dominants is different. The most frequently occurring ants are *Messor* (large carpophages); *Tetramorium* (small carpophages and necrophages); *Lasius* (small necrophages); *Formica* and *Cataglyphis* (large zoonecrophages). However, the characteristic functional groups of these steppes are, as in the woodland belt, some predatory, hunting stratobionts, such as the genera *Myrmica* and *Leptothorax*. The steppes of Kazakhstan, Southern Siberia and Mongolia generally resemble those of Europe, but here the large harvester ants are scarce or totally absent. Reznikova (1975) has shown that cohabitation causes a differentiation of favorable hunting activities among the dominant *Formica pratensis* and *F. uralensis* and the subordinate species, *F. cunicularia* and *F. transcaucasica* of the genus *Formica*. The subordinates, in the presence of dominants, gather food mainly in the grasslands, or, in the absence of grass, on the soil's surface (Reznikova, 1983).

Savannas also possess a structure of ant coadaptive complexes, which is intermediate between that of a desert and that of a tropical forest. Thus, despite the fact that on the well-studied shore of the Ivory Coast (Levieux, 1967, 1973) the total number of species is quite high, only two herpetobionts are truly abundant. These are the large herpetobiont zoonecrophage *Camponotus acvapimensis* and the small herpetobiont zoonecrophage *Acantholepis canescens*. In addition, some geobionts, including those with special food preferences, and some real dendrobionts are found there, as in a tropical forest.

Although at the present time, we already have much information about the structure of multispecies associations and coadaptive complexes, as well as about the influences of the mutual activity of their separate components, the process of evolution by which these complexes were formed remains poorly elucidated. Apparently, researchers are again becoming interested in this question, following a speech by the famous evolutionary biologist and hydrobiologist G. Hutchinson and the theoretical models of May and MacArthur (1972), Brown and Liberman (1973), and Davidson (1976a, b).

Behavioral Adaptations of Desert Ants

At the present time, an abundant literature has accumulated that is devoted to the study of insect adaptation to life in deserts (Buxton, 1923; Pradhan, 1957; Ghilarov, 1964, 1970). However, it is noteworthy that the major part of these works deals with the study of individual "singleton" insects, which possess a whole array of specific morphophysiological adaptations (*Tenebrionidae*; desert roaches; locusts; *Membracidae*, etc.). The social insects, particularly ants, elicited far less attention by investigators, because investigators did not deem such specific adaptations characteristic, inasmuch as many arid land species do not differ at all from species of the same genus inhabiting humid regions (*Camponotus* spp.; many small *Myrmicinae*; the wasps *Vespa* and *Polistes*; and all termite species from Middle Asian deserts). The larvae of social insects inhabiting deserts, like those of humid regions, possess thin and tender skins and do not tolerate humidity at all. That is why, in most reviews, the only other facts mentioned are the light color of some desert ants, the presence of psammophores in certain arid land species and the nocturnal activities of harvester ants.

Delye (1968) made the first serious attempt to review the adaptations of ants to life in deserts. However, even in this work, a traditional emphasis is placed on morphophysiological adaptations, e.g., resistance to high temperature and to high temperature and low humidity. Of all the ethological adaptations, only nest building, the dynamics of activity and food preferences are discussed to some extent. The significance of the diverse morphophysiological and behavioral adaptations to life in deserts by representatives of many trophic groups and species on several levels of social organization is still not fully understood. It is no coincidence that social insects, which depend overwhelmingly on behavioral adaptations, happen to be the most numerous animals in nearly all desert biocenoses. Unlike the vast majority of solitary insects, the desert's conquest takes place principally at the expense of behavioral adaptations in ants (Delye, 1968; Schumacher & Whitford, 1974) and termites (Ghilarov, 1964, 1970).

We have analyzed both morphological and behavioral adaptations of desert ants (Dlussky, 1971, 1972a, b, 1974, 1975a, 1976, 1981b; Dlussky & Saparlyev, 1975). The main behavioral adaptations are: specialized nest construction, storage of food and exploitation of the feeding territory.

The soft-bodied larvae in the nest chambers cannot survive in less than 100% humidity of air in the nest chambers. During the wet season, desert soil is uniformly 100% humid. In the dry season, the soil below 70 cm is completely wet as a result of capillary elevation of underground water. The superficial layer (0–20 cm) is wet during the spring and after dewfall and rain. The layer of soil 20–70 cm is most unfavorable for ant larvae, as it is wet only in the spring and does not become wet during dewfall and rain. At a depth of 3 m or more, salt water is found.

Nest construction of desert ants is connected with the distribution of soil humidity. Most small ants that have no psammophores, e.g., *Plagiolepis, Monomorium, Tetramorium* and *Pheidole* use the superficial layer. *Plagiolepis* construct a complex nest of channels in a depth 1–3 cm under the soil surface to use the water of the dew. More large ants, e.g., *Cataglyphis, Messor variabilis*, and small *Monomorium barbatulum* having specialized psammophores use the third layer. Their nests have a shaft deeper than 70 but usually not more than 120 cm.

Most of the chambers are also concentrated in this layer. Most large ants with developed psammophores, i.e., *Camponotus xerxes*, *Messor aralocaspius*, construct the shaft to underground water and many chambers of their nests are at a depth of some meters under the soil surface.

The most specialized desert ants, e.g., *Camponotus xerxes*, *Messor variabilis* and *Monomorium barbatulum*, construct a special section for the brood, having no entrance to the soil surface but connected at a depth of 70–100 cm by a horizontal channel to the rest of the nest. Only in the desert ants *Cataglyphis oxiana* and *Crematogaster subdentata* we have found specialized chambers and channels for drying of prey under the soil surface.

One of the most important difficulties for desert animals is the variability of resources during the year. Potential food (plants, seeds, insects) are abundant in the spring but scarce at other times. Estimation of potential food abundance is impossible for ants, so only species that store food can inhabit deserts. *Formicinae* and *Dolichoderinae* store liquid food in the crops of plerergates (honey ants). It is not possible for *Myrmicineae* to do this as a result of their proventriculus construction (Eisner & Brown, 1958). Accordingly, most desert *Myrmicinae* are carpophages and store seeds in special chambers of their nests. During the 2 months of the spring, these ants store enough seeds for all the year (Dlussky & Saparlyev, 1975). Storage of dried insects is known only in desert ants *Cataglyphis oxiana* and *Crematogaster* (Dlussky, 1971; 1981b). *C. oxiana* store dried termites and *C. subdentata* dried aphids from desert bushes.

Intensive foraging for short periods of time are necessary for effective exploitation of food resources in the desert. "Mobilization" is defined as any method by which social animals increase the number of conspecifics in their territory. Many mobilization methods are seen in ants in the humid zones, indicating preadaptations in the desert ant. The most primitive method is nonspecific activation (Dlussky, 1975b), which is seen in *Cataglyphis*, *Proformica* and *Acantholepis*. The foragers in these genera find food and activate other foragers but do not give information about the location of the food. *Cataglyphis* collect dead *Messor* ants when they go to their nest during the heat of the day. The *Cataglyphis* forager activates other foragers, each of which goes to its individual feeding territory. This method permits *Cataglpyhis* to quickly collect all dead insects before other animals that are feeding at the same time (Dlussky, 1981a).

Kinoptic mobilization is seen in *Cataglyphis foreli*, a pattern in which vision and moving visual stimuli, without contact, mobilize conspecifics. *C. foreli* prefer to feed on termites (*Anacanthotermes ahngerianus*), which are accessible for very short periods of time when their tubes are destroyed. *C. foreli* are stimulated at first by successful foragers. Upon the second or third return by the foragers to the nest, the other *C. foreli* are kinoptically mobilized to go in the direction from which the foragers have come, rather than to their individual feeding territories.

Most carpophages and nocturnal zoophages are group foragers, another mobilization method. *Pheidole*, *Camponotus* and most of the *Monomorium* use chemical trails, while *Tetramorium* and *M. barbatulum* form chains of individuals that follow scout workers. Chemical trails are more effective, though conservative, than chain formations because in the former method, the trail cannot be changed quickly. However, in the chain method, the place of foraging is more easily changed. The first method, that of chemical trails, is more effective with prey that are large and infrequently found, whereas in the second, the method is more useful for dispersed, abundant and small prey.

An unusual type of mobilization is mass foraging by directionalizing the movement of individuals, seen in desert harvester ants, *Messor aralocaspius* and *M. denticulatus* (Zakharov, 1978; Dlussky, 1981b). These ants collect seeds of ephemeral plants that grow in patches larger than 1 sq. m. They have two diurnal bouts of feeding, during the morning and afternoon. At midday when the soil surface is too hot, they rest. About 3% of the foragers are active foragers (scouts). They have individual feeding territories and are continuously active. Most of the colony workers are passive foragers that have no individual territories and only visit patches of ripe seeds. At the beginning of every feeding bout, crowds of passive foragers gather at the nest entrance. Active foragers run into the crowd and from time to time make short runs in the direction of their feeding territory. Active foragers from plant patches with ripe seeds are more active. Passive foragers follow them at a short distance but then return to the nest entrance. Upon their return, other passive foragers move in the direction from which the returning foragers have come.

All of these directionalized outward and returning movements produce a snakelike general movement of the mass of foragers. The speed of the "snake" is about 15–20 cm/min, whereas the speed of each individual is 1–3 cm/min. This pattern makes it possible to maximize the number of ants recruited and to be independent of chemical trails that would evaporate in the hot soil.

Evolutionary Aspects of Coadaptive Complexes of Desert Ants

The superfamily *Formicoidea*, represented in recent ant fauna by only one single family (Formicidae), first appeared on the boundary of the upper and lower Cretaceous nearby 100 million years ago. Judging by certain morphological characteristics of the earliest *Formicoidea* (Armaniidae), these insects, although resembling today's ants, were not yet truly social (Dlussky, 1983; Dlussky & Fedoseeva, 1987). The most ancient true ant is known from the Paleocene period (Amber of Sakhalin) (Dlussky, 1988). Of the 10 individuals found in Sakhalin, only one belongs to the subfamily *Ponerinae*, and all the others are *Dolichoderinae*. A genus, *Zherichinius*, represented by two species is a typical dendrobiont ant.

Among fossil ant fauna, the most widely studied are those from the Baltic amber (Late Eocene period) (Wheeler, 1915). Almost all present-day subfamilies were found, but representatives of the *Dolichoderinae* (64.3% of deposits) and *Formicinae* (32.8%) were more frequent than any others. Close relationships with recent ant fauna of the Palearctic, the Indo-Malaysian region and Australia have been shown (Wheeler, 1915; Emery, 1920). Although the amber was formed in diverse climatic territories, the overwhelming number of specialized dendrobionts (genera *Gesomyrmex*, *Hypoclinea*, *Oecophylla*, *Tetraponera*) indicates the existence of tropical or subtropical forests. Arid and semiarid regions and their ant faunas were apparently absent during the Eocene. Ant faunas of the tropical or subtropical forests of the Oligocene period are well described by Carpenter (1930) in the sediments of Florisant, California, U.S.A. Here, as in the Baltic amber, the *Dolichoderinae* (63.3% of the imprints), including the dendrobiont *Protazteca* and the *Formicinae* (32.8%) prevail. The weight of the ultrasubordinate *Myrmicinae* (3.7%) is somewhat greater than in the Baltic amber (1.8%). The

Florisant ant fauna resemble also the still undescribed ant from the sediments of the Pacific coast region of Russia.

The ants of the Miocene period are represented chiefly by species of recent genera. The fossil fauna that was found in Dominican amber (Baroni Urbani & Saunders, 1982; Wilson, 1985) was that of tropical forests and similar to recent fauna of Central America. Faunas of Sicilian amber and sediments of Shanwang, South China, (Zhang, 1989) were that of subtropical forests. About a quarter of the ant species that were found in these localities were dendrobionts.

The fossil fauna of the southern part of the temperate climate zone of the Miocene is known to us from findings in sediments of Vishnevaya Balka, Stavropolsky kray, southern Russia (Dlussky, 1981a). The structure of ant complexes of these fauna was quite similar to the structure of ant complexes in the forests of southern Europe. The stratobionts of the genus *Aphaenogaster* (= *Paraphaenogaster*), the *Lasius herpetobionts* and the *Camponotus* and *Liometopum dendrobionts* were the most numerous. The geobionts were represented by two species of *Ponerinae*. The number of specimens of different families was almost the same as those now found in the forests of Southern Europe: *Ponerinae*, 2.7%; *Myrmicinae*, 40%; *Dodlichoderinae*, 4%; *Formicinae*, 53.3% (imprint percentage). The Miocene fauna of Radobon, Yugoslavia, judging by the drawings of Mayr (1867), was similar to the fauna of Vishnevaya Balka, but this cannot be confirmed before further studies are completed.

The most interesting ants of the Miocene period are the ant fauna of sediments of Chon-Tuz (Kirghizia), because they lived in an arid area with a dry climate (Dlusskey, 1981a). Altogether, 22 imprints were found there, usually poorly preserved, which enabled us to describe merely three species: *Ponerites inversa*; *Rhytidoponera kirghizorum*, one of the small representatives of the subgenus *Cholcoponera*, extremely numerous in Australia (evidently stratobionts); and *Kotchorkia*, which is quite large, up to 10 mm long, with a thick skin, well-developed eyes, and normal legs (herpetobionts or primitive dendrobionts). Six of the other imprints belong to species of the genus *Camponotus* (5 males, 9 to 10 mm long, and a single female, 11.5 mm); the other 13 ants are small, poorly preserved, winged individuals, with one joined petiole and a round gaster. We surmise that those ants belong in either the

Dolichoderinae or the *Formicinae* families. One of these imprints (6 mm) looks different from all others. The rest (12 ants) include two uneven groups: 8 ants are 5 mm long, and 4 ants are 7 to 7.5 mm long. They may be males and females of the same ant species, similar to the recent genera *Lasius* or *Iridomyrmex*.

Although there are few extant data, it is possible to infer that the structure of these fossil ant complexes from Chon-Tuz are very different from the ant complexes in today's deserts of Central Asia. First: The carpophages, now prevalent, were then conspicuous by their extreme rarity or total absence. Second: The geobionts and stratobionts, which are now totally absent in deserts, were then abundant. Besides, no dendrobionts were found that lived more recently in areas of tropical forests before these became deserts. The females and workers of dendrobionts, which gnaw passages in wood, have a hard chewing edge along their mandibles, which, as a rule, is well preseved in imprints; often, they have short legs. Specialized species, on the other hand, living in tree crowns have a slender body, long head and limbs. No such forms were found in Chon-Tuz. The ant fauna there resembled most the fauna of a savanna, comparable to the well studied savannas of the Ivory Coast (Levieux, 1967, 1973). There the most numerous species were those of the genus *Camponotus*, among which there were soil surface forms (*C. acvapimensis*, up to 32%) as well as dendrobionts and small herpetobiont zoophages (up to 7%).

The herpetobiont zoophages at Ivory Coast savanna were *Acantholepis* and *Paratrechina*, (up to 17% of all nests). Geobionts and stratobionts were quite numerous, such as *Amblyopone*, *Hypoponera* or *Mesoponera*. Besides, there were many small *Myrmicinae*, which may have been carpophages, but their number was several times smaller than that of zoonecrophages; their diet was not studied. The nest of the *Myrmicinae*, which may be carpophages (*Tetramorium* or *Pheidole*) contributed from 5% to 30% (average, 12.5%) of all nests. Comparatively, ant carpophage nests in the Sahara desert represent 36% to 59% of all nests, (average 52%) (Delye, 1968) and in the Karakrumy desert carpophage nests represent from 58% to 72% of all ant nests (Dlussky, 1981b).

We do not find any real desert complexes or typical desert forms before the uppermost Miocene period. To some extent, this is connected with preservation in amber exuded by ancient trees in tropical, subtropical and possibly even moderate climate forests, whereas the absence or great scarcity of desert ants in the Chon-Tuz sediments, deposited during the Miocene period, can be explained by one or more of the following hypotheses:

1.   The ant genera and species were first formed during the Pliocene period.
2.   The ant genera characteristic of the desert complexes in Central Asia probably did not evolve before the early Miocene period but lived in limited territorial areas and were not as widely distributed as they are today.
3.   The real desert ant species evolved during the Miocene period, although not in Central Asia. They may have migrated there from elsewhere, possibly from a great distance.

Everything that we know about evolution, as shown in ant genera, obviously contradicts the first hypothesis. The male *Paraphaenogaster microphthalmus*, from the Miocene period (Dlussky, 1981a), already had a specialized chest structure of the propodeum, typical of today's members of the genus *Aphaenogaster*, though less vividly expressed. On this basis, we can surmise that males of the genus *Messor*, which have a normal structure of the propedum, and *Aphaenogaster* both diverged earlier than the middle of the Miocene period. As a rule, the ant complexes in deserts are limited to species able to survive only in arid and semiarid regions. If some of them do occasionally venture into humid areas, they can live there only in xerothermal microclimate conditions. They differ from their neighbors, who live permanently in this humid environment, by the latter's slowly evolving preadaptations to future life in savannas and deserts.

The species of the genus *Messor*, apparently derived from *Aphaenogaster*, differ from the latter species by the less specialized structure of the propodeum of the males and by their adaptations connected with their maturation into seed eaters. These workers have a large head, connected with developed mandibular muscles, massive jaws with blunt teeth, a variable worker caste and developed psammophores. Analogous differences are found when we compare the genus Melophorus with other genera of the tribe *Melophorini*; or

Myrmecocystus with other genera of the tribe *Lasiini*; or yet *Pogonomyrmex* with other genera of the tribe *Myrmicini*. Please note that all these genera evolved in arid or semiarid environments. The paleoclimatological data suggest that from the very start of the Cretaceous period, i.e., from the very moment that ants appeared, the dry climate regions were approximately the same as they are today (Sinitsyn, 1967). All the evidence available to us leads us to surmise that all contemporary genera evolved within their respective limits.

As shown in the comparative morphological analysis by Bernard (1973), the species closest to the initial prototypes of the genera *Messor* and *Aphaenogaster* is *M. aphaenogasteroides*, which inhabits the mountains of Afghanistan, and the related species group *M. excursionis*, whose area includes all the deserts of Central Asia, Afghanistan, Kazakhstan and the Middle East. At the same time, the most advanced members of the above mentioned genera live in the Sahara and in Southern Africa. An analogous situation is observed when making a comparative morphological analysis of the tribe *Formicini*. All three species of *Alloformica* inhabit the semisavannas of Middle Asia. Most primitive *Cataglyphis* species (*C. emeryi* and *C. pallida*) are also limited to Southern Kazakhstan, Middle Asia and Afghanistan. As for the genus *Proformica*, the most primitive species is *P. splendida*, whose workers are the least variable and in which the caste plerergaster is absent. This archaic species lives in the high mountain steppes of Tian-Chang. Thus, the combined data of paleontology, comparative morphology and zoogeography testify that the evolutionary formation of ant groups typical of today's Palearctic deserts began early in the Miocene period, somewhere in the vast territory now including the countries named Kazakhstan, Middle Asia and Afghanistan.

The typical desert complexes must have been formed, at least in a yet imperfect way, before the early Miocene, although during the Miocene period, the desert complexes were not as widely distributed as they are today. For example, in Middle Asia, the dominant biomes were still those of savannas. It is logical to suppose that the initial formation of desert ant complexes took place in the poorest, azonal plant communities such as coastal dunes, salt flats or rocky sediments with scant vegetation. By analyzing the ant fauna of Middle Asia's semisavannas, i.e., of territories that are closest in climate to the true savannas of the Miocene period, we see that these territories are inhabited by a great many endemic forms. In Badhkyz and Kopetdag, we encounter *Amblyopone annae, Cerapachys desertorum, Aphaenogaster fabulosa* and *A. messoroides*; in the savannas of Tadjikistan, we find *Aphaenogaster janushevi* and *Alloformica nitidior*, in the valley of Kashkadarya (S. Vzbekistan) we see all the following: *Aphaenogaster raphidiiceps, Alloformica aberrans, Tetramorium striativentre, Leptanilla alexandri* and several others (Dlussky, Sojunov & Zabelin, 1989).

The particular richness of relict archaic forms in the semisavannas of the Kashkadarya valley can be explained by the isolated location of this small territory by mountains. We can surmise that several *Ponerinae, Aphaenogaster, Alloformica* and *Tetramorium*, and also *Camponotus* from subgenus *Tanaemyrmex*, were common and maybe the dominant species of Miocene savannas in the recent territory of Middle Asia and Kazahkstan.

Evidently the earlier adaptive complexes of desert ants originated in isolated azonal arid habitats. One such habitat was sandy dunes, where, as in deserts, the maximal temperatures are high, the diurnal temperature fluctuations are great at the soil's surface, the vegetation is scarce and the subsoil's humus horizon and leaf litter are absent. On the dunes along the river Klyaz'ma, in the Vladimir region, we encounter only three species of ants: *Lasius niger, Formica cinerea* and *Tetramorium caespitum*. The first two species are zoonephrophagous herpetobionts, whose workers greatly differ from each other by their size (worker weights: 2 and 4.5 mg). Furthermore, *F. cinerea* is active exclusively at day, whereas *L. niger* is active both during the day and night. The food habits of *Tetramorium caespitum* are quite varied, but a significant part of its diet consists of seeds of many plant species.

Four common species of ants live on the sandy pasturelands of Northern Japan: *Messor aciculatus, Tetramorium caespitum, Formica japonica* and *Camponotus japonicus*. The first of these species, like all species of the genus *Messor*, is a carpophagous herpetobiont. The three other species feed on various invertebrates, albeit *T. caespitum* switches to a diet of mainly seeds. The average weight of hunted insects of different species are statistically different (Abe, 1971), but the author tells us nothing about the diurnal rhythms of ant activity.

My own summer observations of ants inhabiting the seashore dunes near Hasan reveal that *M. aciculatus* and *F. japonica* are active exclusively in the daytime, whereas the maximal activity of *C. japonicus* occurs at night.

On the seaside dunes of the Southern Pacific coast of Russia (near the city of Hasan), the dominant ant species were *Messor aciculatus* and *Formica transcaucasica*; less often were encountered *F. japonica*, *Lasius alienus* and *Camponotus japonicus*. All these species were herpetobionts; *M. aciculatus* consumed seeds, and the others were zoonecrophages. The latter were characterized by the size of the workers: the median weight of *L. alienus* workers was 1.9 mg; the average weight of *F. transcaucasica* workers was 4.5 mg; that of *F. japonica* workers was 8.6 mg; and that of *C. japonicus* was 25 mg. *F. transcaucasica* and *F. japonica* were active only in the daytime, whereas *C. japonicus* was active mostly at night; *L. alienus* was not studied.

In the Alai valley, at an altitude of 2,600 meters (near Darautkurgan, Kirghizia) on a plateau with cushion-like xerophytes, and with scant grassland vegetation, only three ant species lived. These were the large diurnal zoonecrophage *Cataglyphis aenescens*; the small zoonecrophage, also diurnal, *Proformica coriacea*; and the small carpophage and herpetobiont, *Tetramorium caespitum*. The first two species were very numerous, whereas the third one was rather scarce. On the meadows and steppes around the plateau, all three ant species were also seen, but the dominant forms were several species of *Myrmica* and *Formica lemani*. Encountered, too, were *Formica transcaucasica*, a few specices of *Leptothorax*, *Lasius flavescens* and *Camponotus buddhae*.

The capacity to become adapted to life in deserts is not a specific trait of particular genera: all genera possess this capacity (Kusnezov, 1956). For example, species of the genus *Camponotus* are distributed on all continents, except in the arctic region and in Antarctica. There are desert *Camponotus* specices in the Palearctic region, Africa and Australia. But there are no desert *Camponotus* in North America, despite their presence in humid and semiarid zones, since the Oligocene period. Similarly, species of the genus *Monomorium* inhabit the humid zones of both the Old and New Worlds. However, in Palearctic deserts many of the carpophages are *Monomorium* species, while in the deserts of the New World, they are absent.

Species that form associations in azonal habitats have some similar characteristics. They inhabit practically all types of ecosystems and are very plastic in their behavior. Examples of such species are *Lasius niger* in forest biomes and *Lasius alienus* and *Tetramorium caespitum* in steppe biomes. Their comparatively high level of sociality is combined with the ability to construct nests in a variety of different environmental conditions. They also exhibit a variety of feeding strategies in different environments.

As a rule, several species with similar protein-food spectra live in the humid zone, potentially capable of invading the azonal landscapes with scant vegetation resembling deserts. In fact, however, only one of them, usually the most thermophilic one, does so. Thus, in the forest zone of Russia's Euopean part, four species of *Formica* (*F. cinerea, F. transcaucasica, F. cunicularia* and *F. rufibarbis*) are potentially able to exist on seaside sand dunes, but, in reality, only *F. cinerea* survives there. In Kazakhstan, where *Formica cinerea* is virtually absent, the species that settles in arid environments with scant vegetation is usually *F. cunicularia*, whereas in Tuva (Zhigulskaya, 1968) and on the seaside dunes, where both of these species are virtually absent, the usual settler is *F. transcaucasica*.

It appears that, at the beginning, from the ant fauna then extant, some ant species characteristic of unstable azonal ecocomplexes (seaside areas, dunes, etc.) were formed. Later, after the environment had drastically changed, these specialized ant faunas caught up quite fast with the environmental changes. We surmise that this sequence of events, if not obligatory, is the one occurring most frequently. The first step (or "stage") of the formation of a new desert ant fauna is the appearance of simpler adaptive complexes, based upon the autochthonous evolution of the ant fauna extant in the local humid or semiarid environment.

The earliest ant complexes show the characteristic structure of desert ant complexes, that is, they consist of herpetobiont zoonecrophages that differ from each other by the relationship between their body size and the size of their prey, as well as primitive carpophages. Later the number of species increased as a result of evolution. As desert ants use only one stratum, the soil surface, for feeding, species divergence was based on feeding specialization. Thus, the specificity of the biome and the specificity of the organisms, e.g., ants, determine the structure of the coadaptive complex. In different deserts—Australian, South and North Africa, South and North America, Central Asia—coadaptive complexes of ants consist of species of different genera, but the structure of the complexes is quite the same.

Acknowledgments

I am grateful to Dr. Ethel Tobach for urging me to prepare this chapter and for advice and encouragement during the work. I thank Dr. Joseph Peary for his assistance in the translation of the monograph (Dlussky, G. M., 1981, *Desert Ants*, Nauka Press, Moscow) in which most of the material in this chapter was presented.

References

Abe, T. (1971). The trophic connections of four ant species on sandy pastures. 1. Food and foraging behaviour. *Japan Journal of Ecology, 20*, 219–230 (in Japanese).

Arnoldi, K. V. (1937). The life forms of ants. *Reports of Academy of Sciences of the USSR, 16*, 343–345 (in Russian).

Baroni Urbani, C., & Saunders, J. B. (1982). The fauna of the Dominican Republic amber: The present status of knowledge. *Transactions of the 9th Caribbean Geological Conference, Santo Domingo, Dominican Republic, 1980, Vol. 1*, 213–223.

Bernard, F. (1973). Évolution et biogeographie des Mesor et Cratomyrmex, fourmis moisseneuses de l'ancien monde. *Compte Rendus des Séances de Société de Biogeographie, 437*, 19–32.

Brown, J., & Liberman, M. (1973). Resource utilization and coexistence of seed-eating desert rodents in sand dune habitats. *Ecology, 54*, 775–787.

Buxton, P. A. (1923). Animal life in deserts: a study of the fauna in relation to environment. *Ecology, 54*, 775–757.

Carpenter, F. M. (1930). The fossil ants of North America. *Bulletin of the Museum of Comparative Zoology, 70*, 1–66.

Davidson, D. W. (1976a). Species diversity and community organization in desert seed-eating ants. *Ecology, 58*, 711–724.

Davidson, D. H. (1976b). Foraging ecology and community organization in desert seed-eating ants. *Ecology, 58*, 725–737.

Delye, G. (1968). Recherches sur l'écologie et l'éthologie des fourmis du Sahara. Theses presentées à la Faculté des Science de l'Université d'aux-Marseille pour obtenir le grande de Docteur des Sciences Naturelles. Marseille.

Demchenko, A. V. (1975). Multispecies associations of ants in fir forests in the Moscow region. In *Ants and forest protection. Proceedings of 5th All Union Symposium "Utilisation of ants for the control of forest pests"* pp. 77–82. Moscow (in Russian).

Dlussky, G. M. (1971). The predator ants of Saksaul Forests. In *Ants and forest protection. Proceedings of 4th All Union Symposium "Utilisation of ants for the control of forest pests"* pp. 18–20. Moscow (in Russian).

Dlussky, G. M. (1972a). The behavior of desert ants of the genus *Cataglyphis* when digging into the ground. In V. E. Sokolovv (ed.), *Animal behavior: Ecological and evolutionary aspects* pp. 239–241. Moscow: Nauka Press (in Russian).

Dlussky, G. M. (1972b). The evolution of ant nest construction. *Proceedings of XIII International Entomological Congress* (Moscow, 1969) Vol. 9 (pp. 359–360). Leningrad: Nauka Press.

Dlussky, G. M. (1974). The nest construction of desert ants. *Zoological Journal, 53*,

224–236 (in Russian with English summary).

Dlussky, G. M. (1975a). Ants of the Saksaul Forests of Murgab Delta. In B. M. Mamaev (ed.), *Insects as a component of Saksaul Forest biocenosis* (pp. 159–185). Moscow: Nauka Press (in Russian).

Dlussky, G. M. (1975b). Nonspecific activation of scouts in Myrmica. In *Ants and forest protection. Proceedings of 5th All Union Symposium "Utilisation of ants for the control of forest pests"* pp. 125–133. Moscow (in Russian).

Dlussky, G. M. (1976). The organizing of group foraging by ant *Pheidole pallidula*. In V. E. Sokolov (ed.), *The group behavior of animals* pp. 97–99. Moscow: Nauka Press (in Russian).

Dlussky, G. M. (1981a). Ants of Miocene (Hymenoptera, Formicidae) from USSR. In V.N. Vishnjakova, G. M. Dlussky & L. N. Pritykina (eds.), *New fossil insects from the territory of the USSR. Proceedings of Paleontological Institute of the USSR Academy of Sciences*, Vol. 183 (pp. 64–83). Moscow (in Russian).

Dlussky, G. M. (1981b). *Desert ants*. Moscow: Nauka Press (in Russian).

Dlussky, G. M. (1983). The new family Late-Cretaceous Hymenoptera—an "intermediate link" between ants and Scolioidea. *Paleontological Journal, 3,* 65–78 (in Russian).

Dlussky, G. M. (1988). Ants of Sakhalin amber (Palaeocene). *Paleontological Journal, 1,* 50–61 (in Russian).

Dlussky, G. M., & Fedoseeva, E. B. (1987). The origin and early evolutional steps of ants (Hymenoptera: Formidicae). In A. G. Ponomarenko (ed.), *Crecaceous biocenotic crisis and the evolution of insects* pp. 70–144. Moscow: Nauka Press (in Russian).

Dlussky, G. M., & Saparlyev, K. (1975). Activity dynamics of desert harvester ants. *Ecologie (Sverdlovsk), 4,* 79–85 (in Russian).

Dlussky, G. M., Sojunov, O. S., & Zabelin, S. I. (1989). *Ants of Turkmenistan*. Ashkhabad: Ylym Press (in Russian).

Eisner, T., & Brown, W. L. (1958). The evolution and social significance of the ant proventriculus. *Proceedings of the Tenth International Congress of Entomology* (Montreal, 1956), Vol. 2, 503–508.

Emery, C. (1891). Le formiche dell'Ambra Siciliana nel Museo Mineralogico dell'Università di Bologna. *Memoire della Reale Accademia delle Scienz dell'Instituto di Bologna*, Ser. 5, Vol. 1, 567–594.

Emery, C. (1920). La distribuzione geographica attuale della formiche. *Memoire della Reale Accademia Lincei* (Roma), Ser. 5a. *Classe di scienze fisiche matematiche e naturali.* Vol. 13, 357–450.

Ghilarov, M. S. (1964). The adaptation of insects to life in deserts. *Zoological Journal, 43,* 443–454 (in Russian with English summary).

Ghilarov, M. S. (1970). *Adaptation of Arthropoda to terrestrial life.* Moscow: Nauka Press (in Russian).

Kaczmarek, W. (1953). Investigations of forest ants complexes. *Ecologia Polska, 1,* 69–96 (in Polish).

Kashkarov, D. N., & Korovin, E. P. (1936). *Life of deserts.* Moscow-Leningrad: Biomedgiz Press (in Russian).

Kusnezov, N. (1956). A comparative study of ants in desert regions of Central Asia and of South America. *American Naturalist, 90,* 349–360.

Levieux, J. (1967). Étude de peuplement en fourmis tericoles d'une savanne préforestière de Côte d'Ivoire. *Revue d'Écologie et Biologie du Sol, 10,* 379–428.

Levieux, J. (1973). Étude de peuplement en fourmis tericoles d'une savage préforestière de Côte d'Ivoire. *Revue Écologique et Biologie Soleil, 10,* 379–428.

May, R. M., & McArthur, R. H. (1972) Niche overlap as a function of environmental variability. *Proceedings of the National Academy of Science, 69,* 1109–1113.

Mayr, G. L. (1867). Vorläufige Studien über die Radoboj-Formiciden, in der Sammlung der geologischen Reichsanstalt. *Jahrbuch der geologischen Reichsanstalt, Wien, 17,* 47–62.

McNab, B. (1971). The structure of tropical bat faunas. *Ecology, 52,* 352–358.

Petrov, M. P. (1973). *The deserts of the earth.* Mosow: Nauka Press (in Russian).

Pradhan, S. (1957). The ecology of arid zone insects (excluding locusts and grasshoppers). In *Arid Zone Researches. Human and Animal Ecology* (pp. 199–240). UNESCO: Paris.

Rezinkova, Z. I. (1975). Nonantagonistic mutual relationships of ants with similar ecological niches. *Zoological Journal, 54,* 1021–1031 (in Russian with English summary).

Reznikova, Z. I. (1976). Hierarchy of species in the complexes of steppe ants. In V. E. Sokolov (ed.), *The group behavior of animals* (pp. 315–318). Moscow: Nauka Press.

Reznikova, Z. I. (1983). *Interspecific relations between ants.* Novosibirsk: Nauka Press (in Russian).

Schumacher, A., & Whitford, W. G. (1974). The foraging ecology of two species of Chihuahuan desert ants: *Formica perpilosa* and *Trachymyrmex smithi neomexicanus* (Hymenoptera: Formicidae). *Insectes Sociaux, 21,* 317–330.

Seyma, F. A. (1972). Spatial distribution of ant workers in the biocenoses. *Zoological Journal, 51,* 1322–1328 (in Russian with English summary).

Sinitsyn, V. M. (1967). *Introduction to paleoclimatology.* Moscow: Nedra.

Stebaev, I. V. (1971). The structure of *Formica pratensis* territory, and interspecies and intraspecies relationships of ants. *Zoological Journal, 50,* 1504–1519 (in Russian with English summary).

Stebaev, I. V., Yamkovoy, V. N., Stepanova, V. V. Milovidova, I. M., & Tarasenko, A. V. (1967). Some interspecies and intraspecific relationships of some species of the ant genus Formica in the steppes of South Siberia. In *Ants and forest protection. Proceedings of 3rd All Union Symposium "Utilization of ants for the control of forest pests"* (pp. 45–48). Moscow (in Russian).

Wheeler, W. M. (1915). The Ants of the Baltic Amber. , 1–142.

Wilson, E. O. (1985). Ants of the Dominican amber (Hymenoptera: Formicidae). *Psyche, 92,* 1–37.

Zahkarov, A. A. (1978). *Ant, Family, Colony.* Moscow: Nauka Press (in Russian).

Zahkarov, A. A. (1991). *The Organising Ant Societies.* Moscow: Nauka Press (in Russian).

Zhang, Jan Feng (1989). *Fossil Insects from Shanwang, Shandong, China.* Junan, China: Shandong Science and Technology Publishing House (in Chinese with English summary).

Zhigulskaya, Z. A. (1968). The population of ants (Formicidae) in the steppes of Tuva. In *Animal populations of soils in woodless ecosystems of the Altay-Sayan Mountain System* (pp. 115–139). Novosibirsk (in Russian).

Section II

Social Organization in Ants

Task Allocation and Interaction Rates in Social Insect Colonies

Deborah M. Gordon

In a social insect colony, various tasks are accomplished. These include foraging for food and water, nest construction and repair, cleaning and brood care. Ideas about division of labor, and about task allocation, represent two different approaches to finding out how these tasks are organized. The "division of labor" approach is based on the idea that individuals are specialized to certain tasks. This approach raises questions about the factors that determine which individual performs which task. The study of task allocation, however, examines the factors that determine how many individuals perform each task, regardless of those individuals' identities. This second approach raises questions about how much of a task gets done and how this depends on external conditions (Gordon, 1996).

Adam Smith, the Scottish economist, first used the phrase "division of labor" in his 1776 book, *The Wealth of Nations*. Since then, the assembly line has become the most familiar example of division of labor: each individual worker on the assembly line does some part of a larger task, and each individual does the same task over and over. In the social insect literature, "caste," which was first used to distinguish reproductive from nonreproductive females, came to mean a group of individuals that share a tendency to perform some task. In some species, individuals were thought to perform some task throughout their lives; in most others, individuals specialize in one task for some period of time, then become more likely to perform other tasks. This shift in individual specialization, in the course of discrete periods of weeks or months over an individual's lifetime, is known as "age polyethism."

The study of division of labor in social insect colonies was linked to a set of evolutionary questions. If a colony consists of groups of specialized individuals, and each individual invariably performs the same task, then the amount of the task to be accomplished is determined by the number of individuals in each task group. According to this argument, if there is natural selection on the amount of work that is done, there will be natural selection on numbers of individuals to do each task. For example, if natural selection favors colonies that do more foraging, then natural selection would favor colonies that contain more specialized foragers. Oster and Wilson (1978) proposed the concept of the "adaptive caste distribution"; that is, the distribution of workers in each task group that generates the optimal performance of each task.

If all individuals were specialized to perform some task, and each individual continuously repeated the same task, then task allocation would be fully determined by division of labor. But recent data shows that individuals are not all specialized. Moreover, even if individuals were specialized, individuals are often inactive. That is, even if an individual would tend to perform only one task if it were active, it is not active all the time. Thus the effort, in numbers of workers allocated to each task depends in part on how many individuals actively pursue a given task.

One example of data showing the distinction between division of labor and task allocation comes from a series of perturbation experiments performed with harvester ant colonies. Each experiment led to increases or decreases in numbers of workers engaged in a single task. Only the workers engaged in one task were directly affected by the perturbations. But the results showed that when numbers engaged in one task change, numbers engaged in another are altered as well (Gordon, 1986, 1987). Experiments with marked individuals showed that when environmental conditions change, individuals switch tasks (Gordon, 1989b). In addition, when environmental conditions change, individuals alter their activity levels (Gordon, 1986, 1987).

These results raise a new set of evolutionary questions (Gordon, 1989a). If natural selection acts on the results of a colony's work, then it may act on the dynamics that determine task allocation. For example, suppose natural selection favors colonies that acquire more food. The numbers of ants foraging will depend on the dynamics of task switching: under what conditions will individuals switch tasks to forage? Numbers of ants foraging will also depend on the dynamics that determine when a forager decides to forage actively, rather than waiting inside the nest. Understanding the evolution of colony organization requires us to understand these dynamics.

How Does Task Allocation Operate?
Models of Colony Organization

Colonies make complex adjustments to changes in their environments. When a new food source appears, ants engaged in other tasks switch tasks to forage. Ants that were previously inactive come out of the nest to retrieve the food. When there is a great deal of nest maintenance work to be done, workers that previously did brood care come out of the nest to help. Foragers remain inside the nest while the nest maintenance crew goes to work. But these complex adjustments result from the behavior of individuals. Individual ants are not capable of complex decisions. No ant can count the number of active nest maintenance workers and decide whether to forage accordingly. No ant directs the behavior of others; the "queen" has no authority. She merely stays inside the nest laying eggs. Understanding how task allocation operates means finding out how individuals, making decisions based on local information, work together to produce the complex behavior of colonies.

Individual decisions involved in task allocation have two consequences: individuals decide which task to perform, and individuals decide whether to be active or inactive. What cues do individuals use in making these decisions? The environment is one source of information. An individual may decide to forage when it comes upon food. Other individuals may also provide information. An individual may decide to forage as a result of an encounter with another individual whose behavior or chemical state indicates the presence of food.

There are several lines of evidence that cues from other individuals, as well as cues from the environment, influence individual decisions about task allocation. First, in harvester ants, ants engaged in one task, such as foraging, rarely contact environmental stimuli associated with another task, such as nest maintenance. Foragers travel to different places outside the nest than do nest maintenance workers. But each individual makes a series of trips in and out of the nest. A forager may come back into the nest, drop its seed, and then turn and go back out on another foraging trip. A nest maintenance worker may come out of the nest carrying a bit of dirt, put it down, then go back into the nest and pick up another bit of dirt. In the upper chambers of the nest, workers of different task groups mingle together. It is easier to see how ants engaged in one task might acquire information from ants engaged in another task than to see how they would acquire it directly from a patch of the environment they rarely visit.

A second line of evidence comes from theoretical models of task allocation. One such model is a neural network (Gordon, Goodwin & Trainor, 1992), applied here by drawing on the analogy between ants and brains. In both kinds of systems, the units (ants or neurons) engage in simple behavior, using only local information—but the whole system (colonies or brains) makes complex adjustments to environmental variation. In a neural network, the units' simple binary decision (off or on) depend only on interactions with other units. In our model, we represented the tasks performed outside the nest by 4 possible states, corresponding to midden work, foraging, nest maintenance and patrolling. Each of these could be either active or inactive, leading to a total of 8 states (4 tasks, active or inactive). Which state an individual takes depends only on the sum of weighted interactions among individuals. That is, in this model, only interaction with others determines the results—the only information from the environment is the starting condition of numbers of ants in each of the 8 states. Using only rules based on interactions, we were able to simulate some of the empirical results: a change in numbers engaged in one task led to changes engaged in other tasks. This theoretical work does not demonstrate that real ant colonies behave like neural networks. It does show, however, that simple cues like numbers of interactions can be sufficient to generate one aspect of colony-level task allocation: numbers of ants performing one task influences the numbers performing another.

In a more recent model of task allocation (Pacala, Gordon & Godfray, 1996), an ant uses both environmental stimuli and social interaction in its decisions whether to be active and which task to perform. We assumed that an ant is likely to switch to do task i

when it encounters another ant that has successfully performed task *i*, and that it will quit task *i* after a certain amount of time spent unsuccessfully attempting to perform task *i*. For example, a foraging ant would be successful, and able to cause other ants it meets to become foragers, when it finds food; it would quit foraging after a certain amount of time spent searching for food without finding any. The dynamics of this model are interesting. As might be expected, at equilibrium such a system reaches the ideal free distribution, where all task groups, each consisting of the ants performing a particular task, experience the same level of success.

This model shows how the efficiency of task allocation may depend on colony size. As the system approaches equilibrium, larger colonies track the environment more quickly. This effect is a consequence of the contribution of social interaction to task allocation. It arises from the way interaction rates depend on colony size. The simplest null model is that ants, like particles in Brownian motion, collide at random. In that case the interaction rate is the square of the number of ants. The relation of interaction rate and colony size depends on the details of how ants move around and how they respond to encounters (Adler & Gordon, 1992). But in general, it is reasonable to expect numbers of encounters to be proportional to numbers of ants. In large colonies, with high interaction rates, successful ants are more likely to encounter unsuccessful or inactive ones. This means that large colonies approach the equilibrium distribution of ants into tasks more quickly than small colonies.

However, there is a disadvantage to large colony size. A successful ant may encounter more ants, and encourage more ants to become active, than the environment really warrants. This is a disadvantage if the ants are better off remaining inactive when there is nothing they can usefully accomplish. For example, a successful forager may come back to the nest and, in a very large colony, encounter hordes of inactive ones. These will be stimulated to search for food, even if there is no food to be found. In a smaller colony, the quitting rate of unsuccessful foragers could correct for this. One solution to this problem would be for large colonies to curtail the numbers of encounters at very high densities.

These theoretical results suggest that interaction rates may be important cues to individual decisions about task allocation. But empirical work is needed to discover whether, and how, real ants behave like the ants in these models. I have begun to examine whether ants might use the rate of antennal contact as a cue to worker density. Suppose an ant has a certain threshold time interval at which it expects to meet a nestmate. If this threshold is reached, the ant continues as it is. If it meets ants too often, or fails to meet them often enough, it changes its behavior.

This idea interests me because it is an example of how a pattern of group behavior might determine an individual's experience, though the individual need not be aware of the group's pattern. How often ants meet will depend on how they move around and on how many ants there are participating in the group. Every ant's behavior contributes to the probability of encounter for every other ant, so the rate of interaction each individual experiences is a consequence of the number of ants in the whole group. Yet each ant only needs to keep track of how many ants it meets, or the rate at which it meets other ants, and it will be able to monitor changes in the density of the whole group.

Empirical Studies of Encounter Patterns

Antennal contact is a pervasive element of the behavior of ants. Many of the species that come into buildings and nest in pavement—the species that most people see—form trails. If you watch ants moving back and forth along a trail, you will see them stop to touch antennae with ants coming in the other direction. Ants are very sensitive to chemical information and use their antennae to perceive smell. In the course of antennal contact, an ant can tell whether another is a nestmate. Antennae are used in many elaborate, specific, behavioral rituals. But the function of the ordinary and widespread brief antennal contact is not understood.

Ant species differ in patterns of antennal contact and in the contexts in which antennal contact is used. The fire ant, *Solenopsis invicta*, seems to use antennal contact to signal the presence of food (Gordon, 1988). When fire ants explore a novel region, they do so in a characteristic temporal sequence. Some ants seem to act as sentries, coming into the new space and standing still. Others move through the new area, engaging in antennal contact with slower-moving ants and the stationary sentries. When food is discovered in a novel region, the rate of antennal contact increases. These different path shapes are characteristic of distinct groups of ants, each of which corresponds to a particular head width and body size. Further studies of contact rates showed that in *S. invicta*, the path shape tends to be associated with the presence of food rather than other environmental factors such as the location of the nest or time of day. Other species use contact rate differently (Gordon, Paul & Thorpe, 1993). For example, the ant *Lasius fuliginosus* lives in a stable environment; colonies nest under large oak trees and persist for many years. Foragers collect honeydew from aphids in trees, so foraging trails and aphid associations last for many years. In this species, contact rate is stable, regardless of location or time of day.

To investigate how contact rate depends on density, ants were kept in flat, featureless arenas. The experiment involved 3 group sizes of ants and 3 arena sizes. Contact rate was monitored by observers recording the numbers of ants in each square of a grid underlying the arena. The results showed that contact rate is not the result of random collisions (Gordon et al., 1993). Instead, at large densities, colonies regulated contact rate. Instead of increasing quadratically with number of ants, it levelled off at high densities. Ants appear to avoid contact when they are very crowded; as a colleague put it, they "just say no to contact." This result is exactly the one predicted by the dynamical model described above. If individual decisions about task allocation result from cues affected by interaction rate, then colonies should suppress contact rate at high densities. Of course, this does not show that task allocation operates in the way described by the model. It does, however, encourage us to continue investigating how contact rate contributes to task allocation.

How often ants meet depends on the details of how they move around. How ants move around varies, depending on what the ants are doing. For one task, patrolling, the functional importance of path shape is obvious. Patrolling ants monitor the colony's environment. In harvester ants, patrolling is done at a particular stage in the day's activity period and performed by a distinct group of workers. It is the patrollers that search for food and respond when there is a disturbance involving alien ants or the destruction of the colony's nest or nest mound (Gordon, 1983, 1987). Patrollers are active early each morning, before the foragers become active. Foragers travel in the direction chosen by the patrollers to search for food (Gordon, 1991). If patrollers encounter a new food source, they will recruit to it. Later on, if foragers encounter a new food source, they will walk right by it to reach the searching region earlier indicated by the patrollers. In harvester ants, searching for seeds on the ground in a location the colony may occupy for 15–20 years, patrolling once a day may provide the colony with all the information it needs about the environment. By contrast, patrolling in the invasive Argentine ant, *Linepithema humile*, is perpetual and intensive (personal observations). The Argentine ant forms "supercolonies," diffuse agglomerations of queens and workers that bud off and spread to new locations. When ants find food or water, they will form a trail to it. Ants that arrive at the source will immediately begin to search the area for another food soure. This leads to systems of long trails that can be quickly modified at any time, as resources are depleted and new ones are found.

Patrolling behavior varies widely among ant species, and this variation is linked to ecological differences among species, especially how their resources are distributed in time and space. But there are some constant themes in the relation between path shape and the effectiveness of patrolling. Since ants respond to chemical cues, an ant must get close to an event in order to discover it. The best way for a colony to ensure finding any relevant event, such as food or the presence of an enemy, would be to have an ant everywhere all the time. But since ant numbers are limited, the ants have to move around in a way that puts an ant almost everywhere almost all the time. So one element of efficient patrolling behavior is to cover the ground broadly enough, and often enough, to find relevant events. How much ground they have to cover, and how often, will depend on how events are scattered, how often they appear and how often they last. There is a second important element of patrolling behavior. To transfer information, for example through contact, or by depositing a chemical cue on the ground or into the air, ants must be close together. Patrolling ants need to keep close enough together to be able to inform each other if something happens.

In a model of patrolling behavior in ants, Adler and Gordon (1992) showed how searching efficiency depends on path shape and numbers of ants. Ants patrolled a region with a slightly repellent boundary, in which events appeared at randomly selected locations. We considered the case in which the duration of events was random, and the case in which all events were of equal duration. Searching efficiency was measured as the proportion of all events that were discovered by some ant. Path shape can be characterized in terms of the turning angle, the angle between successive steps or segments of the path. When the variation in turning angle is high, the path is very wiggly in shape, culminating in a completely random walk. When the variation in turning angle is nonexistent, the path is a straight line or perfect circle; every step is at the same angle from the previous one. The model shows that when numbers of ants are low, a very wiggly path is inefficient because ants tend to get stuck in one place and search that same place over and over. But as numbers of ants increase, this cost of a wiggly path becomes less important, because the ants are sufficiently crowded that if one ant doesn't find something, another ant will.

I examined the relation of path shape and numbers of ants in Argentine ants (Gordon, 1995). I chose this species because their ability to invade existing ant communities, and their rapidity at finding new sources of food and water, suggest they are very efficient patrollers. The results are consistent with the predictions of the model described above. These experiments show that ants can regulate both encounter rate and path shape in response to density. Thus ants are capable of the delicate adjustments necessary if they are to use interaction rates in task allocation. However, we are still far from showing that ants do use interaction rates in task allocation. Experiments are underway to investigate this possibility.

Evolution of the Dynamics of Task Allocation

Understanding how task allocation operates is only a step toward understanding the evolution of colony organization. Our interest in task allocation rests on the premise that a colony's proficiency at performing its tasks is related to colony fitness. This relation remains to be demonstrated. Do colonies vary in the amount of work they get done? If so, is this variation related to reproductive success? These two questions must be answered before the mechanisms of task allocation become relevant to the evolution of behavior. If task allocation is a component of colony fitness, then we need to know the details that determine how individuals decide which task to perform, and whether to be active, to know why one colony does better than another. Then we will be justified in speculating about why one set of behavioral dynamics may be more efficient than another. What is needed, then, is more work on the population biology of social insects, as well as more work on the fine-tuning of individual and colony behavior.

References

Adler, F. R., & Gordon, D. M. (1992). Information collection and spread by networks of patrolling ants. *American Naturalist, 40,* 373–400.

Gordon, D. M. (1983). The relation of recruitment rate to activity rhythms in the harvester ant, *Pogonomyrmex barbatus. Journal of the Kansas Entomological Society, 56(3),* 277–285.

Gordon, D. M. (1986). The dynamics of the daily round of the harvester ant colony. *Animal Behaviour, 34,* 1402–1419.

Gordon, D. M. (1987). Group-level dynamics in harvester ants: young colonies and the role of patrolling. *Animal Behaviour, 35,* 833–843.

Gordon, D. M. (1988). Group-level exploration tactics in fire ants. *Behaviour, 104,* 162–175.

Gordon, D. M. (1989a). Caste and change in social insects. In P. Harvey and L. Partridge (eds.), *Oxford surveys in evolutionary biology,* Vol. 6 (pp 56–72). Oxford: Oxford University Press.

Gordon, D. M. (1989b). Dynamics of task switching in harvester ants. *Animal Behaviour, 38,* 194–204.

Gordon, D. M. (1991). Behavioral flexibility and the foraging ecology of seed-eating ants. *American Naturalist, 138,* 379–411.

Gordon, D. M. (1995). The expandable network of ant exploration. *Animal Behaviour, 50,* 995–1007.

Gordon, D. M. (1996). The organization of work in social insect colonies. *Nature,* 380, 121–124.

Gordon, D. M., Goodwin, B., & Trainor, L. E. H. (1992). A parallel distributed model of ant colony behaviour. *Journal of Theoretical Biology, 156,* 293–307.

Gordon, D. M., Paul, R. E. H., & Thorpe, K. (1993). What is the function of encounter patterns in ant colonies? *Animal Behaviour, 45,* 1083–1100.

Oster, G., & Wilson, E.O. (1978). *Caste and ecology in the social insects.* Princeton: Princeton University Press.

Pacala, S. W., D. M. Gordon, & H. C. J. Godfray. (1996). Effects of social group size on information transfer and task allocation. *Evolutionary Ecology, 10,* 127–1007.

Division of Labor in Social Insect Colonies

Self-Organization and Recent Revelations Regarding Age, Size and Genetic Differences

Robin J. Stuart

Sociality has been an enormously successful strategy for insects. Indeed, the modern insect fauna is predominantly social with the social bees, wasps, ants and termites constituting over 75% of the total insect biomass. In an Amazonian rain forest, about a third of the entire animal biomass is composed of ants and termites, with each hectare of soil containing in excess of 8 million ants and 1 million termites. Social insects are also thought to dominate in most other habitats around the world (Hölldobler & Wilson, 1990; Wilson, 1985). It has been suggested that one of the primary reasons for the immense success of social insects is that social evolution has enabled the closely related individuals that typically comprise a colony to effectively divide up the tasks necessary for survival and reproduction and to produce specialists that effectively increase the colony's overall efficiency in comparison to solitary insects (Jeanne, 1986a, b; Oster & Wilson, 1978; Wilson, 1985).

The most fundamental division of labor that occurs within a social insect colony is between the queen and the workers; a dichotomy that separates reproductive and nonreproductive tasks. However, there is often considerable division of labor within the worker caste as well, with different workers showing various degrees of specialization on different tasks. Usually division of labor among workers (or polyethism) is correlated with the age, size or shape of the workers involved (Hölldobler & Wilson, 1990; Lenoir, 1987; Page & Robinson, 1991; Sudd, 1982). However, these factors alone are often insufficient to explain the full range of variability in the behavior of workers that is observed within and among social insect colonies, and much current research is directed toward documenting this variability and determining the factors and mechanisms that are responsible (Calabi, 1988; Gordon, 1989a; Page & Robinson, 1991; Tofts & Franks, 1992). In the process, researchers have begun to question and to modify previous models of polyethism and to offer new ideas regarding the functional and evolutionary dynamics of social organization in insect colonies. For example, the concept of self-organization in complex systems is providing an important new approach to explaining certain aspects of division of labor (Deneubourg, Goss, Pasteels, Fresneau & Lachaud, 1987; Gordon, Goodwin & Trainor, 1992; Page & Mitchell, 1990).

The present paper discusses some current ideas regarding division of labor among workers in social insect colonies and the evolution of worker caste systems. I will review some of the more general aspects of division of labor in social insects and focus on some recent results of special interest. I will deal primarily with ants, since that is my particular area of research, but I will also make numerous references to honeybees because of the extensive and detailed recent work that has been conducted on this subject in this particular species. The ability to breed honeybees using artificial insemination has made them especially useful for the study of genetic components of division of labor, and the demonstration of genetic effects constitutes an important recent advance in our knowledge of polyethism (Page & Robinson, 1991). Indeed, genetic effects must now take their place along with developmental and environmental effects as one of the major classes of influences on division of labor in social insects. However, to date, research on genetic components of division of labor has been conducted almost exclusively with honeybees, and the extent to which these results apply to other social insects remains largely undetermined, except for the work of Carlin, Reeve and Cover (1993), Snyder (1992) and Stuart and Page (1991). Overall, studies of division of labor have involved relatively few species, but generalizations within and among groups have been common and could mask a good deal of diversity. Much more research remains to be conducted, and it is likely that many new discoveries will ensue.

Age Polyethism, Size Polyethism and a Revised Definition of the Monomorphism-Polymorphism Boundary

Traditionally, division of labor among workers in social insect colonies has been considered to be of two fundamental types: age polyethism and size polyethism. Age polyethism is widespread, and as workers age they typically progress from performing certain tasks within the nest (e.g., brood care) to riskier outside tasks (e.g., foraging). Size polyethism occurs more rarely and involves the differential performance of tasks by workers of different sizes, workers that are often distinctly polymorphic with striking differences in the size and shape of various body parts. Often the largest ant workers (or majors) have disproportionately large heads and function primarily as soldiers, as seed grinders or for food storage, while the smaller workers (or minors) perform more general tasks (Hölldobler & Wilson, 1990; Lenoir, 1987; Page & Robinson, 1991; Sudd, 1982). Age polyethism has been studied extensively in the honeybee, and we have much information on the various factors that influence or regulate age polyethism in this particular species (Page & Robinson, 1991).

Age polyethism is also characteristic of most ant societies (Hölldobler & Wilson, 1990; Lenoir, 1987; Sudd, 1982) with only one species, the primitive ponerine ant, *Amblyopone pallipes*, having been shown to have no age polyethism (Traniello, 1978). However, in contrast to honeybees, some ant species show considerable size variability, and size can be a very important aspect of division of labor in ant colonies. A prime example of size variability occurs in the Asian marauder ant, *Pheidologeton diversus*, which exhibits the greatest size variability known among ants with the largest and smallest workers differing in size by about 10 times and in dry weight by about 500 times. Not surprisingly, division of labor in this species has both age and size components (Hölldobler & Wilson, 1990; Moffett, 1988).

In typical age polyethism, the probability that a worker will perform a particular task is correlated with its age. As a worker ages, it can pass through various phases during which it performs different groups of tasks. Workers that are in the same phase are said to represent a particular temporal caste. This progression is often associated with changes in the development of various exocrine glands that are necessary for the performance of particular tasks. Analyses of age polyethism in the honeybee and in the ant, *Pheidole dentata*, suggest that the honeybee has four temporal castes while this particular ant species has three. In discrete caste systems such as these, workers progress through well-defined temporal castes, and the probability peaks for various tasks within the same group are tightly synchronized or concordant. When the probability peaks are not tightly synchronized and discrete temporal castes cannot be resolved, the caste system is referred to as continuous. Examples of both discrete and continuous caste systems are known, but the relative pervasiveness of these patterns among species is unclear (Hölldobler & Wilson, 1990; Page & Robinson, 1991).

Workers in colonies of most social insect species are very uniform in size and are referred to as monomorphic. However, in some species, workers vary considerably in size and are considered polymorphic. For a more precise definition, it is necessary to analyze the worker size-frequency distribution within colonies, and to consider the developmental basis of size variability among workers. Wilson and colleagues (Hölldobler & Wilson, 1990; Oster & Wilson, 1978; Wilson, 1968) defined monomorphic species as those with workers that were either isometric (i.e., exhibited no differential growth of body parts with increasing size) or displayed very little size variability or both; and polymorphic species as those with workers that were nonisometric over a sufficient range of worker sizes to produce individuals of distinctly different proportions.

Recently, however, Wheeler (1991) has suggested that these definitions be revised. Her studies indicate that the number of physical castes within an ant species is best defined by the number of sets of growth rules or developmental programs that are necessary to produce the worker caste system for that species, and that polymorphism can involve alterations in the size-frequency distribution but be associated with little or no allometry (i.e., little or no differential or nonisometric growth). According to Wheeler, monomorphic species produce a unimodal distribution of worker sizes that tends to be narrow and symmetrical because development involves approximately the same critical size for the induction of pupation and the same growth parameters for all individuals.

However, allometry can vary among monomorphic species and depends on the growth parameters involved. The simplist polymorphisms can involve a reprogramming of critical size for the induction of pupation during larval development so that there are two sets of critical sizes and one set of growth parameters. This results in a skewing of the size-frequency distribution toward bimodality with the separation of the modes and the proportion of individuals within each mode depending on the various factors that might influence the setting of critical size and the sensitivity of larvae to the relevant cues. More complex polymorphisms can result from reprogramming both critical size and growth parameters so that the allometric relationships of workers within each mode of the frequency distribution also differ. Increasingly complex polymorphisms are possible by appending revised programs to the ends of existing developmental pathways. By plotting the worker size-frequency distribution together with log-log plots of the relationship between the sizes of various body parts (e.g., head width versus alitrunk width) to indicate the degree of allometry, a fairly good picture can be obtained of the degree of polymorphism of a particular species.

As indicated above, the worker size-frequency distribution within colonies is important for determining the degree of polymorphism exhibited by a species, but this distribution must be interpreted with caution since it might include individuals produced over a considerable period of time and might be influenced by environmental variability and differential survivorship according to size (Wheeler, 1991). Indeed, larval nutrition is thought to have an important impact on worker size across seasons or brood cohorts, a phenomenon that sometimes produces clearly bimodal worker size-frequency distributions within colonies even though individual age cohorts are monomorphic (Brian & Brian, 1951; Delage-Darchen, 1974; Gray, 1971; Markin & Dillier, 1971; Plateaux, 1970, 1971; Rissing, 1987; Stuart, in press). Thus, to determine the fundamental size distribution of workers being produced by a colony, it is important to examine the size-frequency distribution of workers that develop during a very brief period of time, perhaps by assessing the sizes of worker pupae that are simultaneously present in the colony (e.g., Wheeler, 1991) or of workers that are produced within a single brood cycle (e.g., Stuart, in press). Nonetheless, changes in worker sizes produced by colonies over time can also be an important aspect of colony ontogeny and of division of labor (Tschinkel, 1988; Wilson, 1983).

Most ant species appear to be strictly monomorphic, but a sizeable minority are polymorphic to varying degrees, with worker size-frequency distributions that exhibit two, three or even four peaks. In some species there are obvious discontinuities or breaks in the size-frequency distribution such that various classes of major and minor workers are clearly distinguishable. As a rule, the greater the size variability and allometry in a species, the more pronounced the division of labor among the physical castes. Moreover, as physical castes age, they can also exhibit systematic changes in behavior such that each physical or size caste can effectively contain a series of age or temporal castes and, thereby, contribute to even more complex patterns of division of labor within colonies (Hölldobler & Wilson, 1990; Wheeler, 1991).

Do Monomorphic Ants Exhibit Size Polyethism?

According to Hölldobler & Wilson (1990), most monomorphic ant species display no apparent size bias in division of labor. However, few studies appear to have systematically examined this question, and reports of size polyethism in monomorphic ants involve a variety of species including the formicine *Myrmecocystis mimicus* (Hölldobler, 1981; Hölldobler & Wilson, 1990), the myrmicine *Leptothorax longispinosus* (Herbers & Cunningham, 1983) and the ponerine *Paraponera clavata* (Breed & Harrison, 1988). I recently reexamined division of labor in *L. longispinosus* (Stuart, in press) and, contrary to Herbers and Cunningham (1983), found that size biases with respect to division of labor in this species were inconsistent among colonies and were, therefore, not indicative of size polyethism. Moreover, my results also cast serious doubts on other reports of size polyethism in monomorphic ants. The evidence is considered below.

Herbers and Cunningham (1983) analyzed the behavioral repertoire of workers in 4 *L. longispinosus* colonies and found that division of labor was characterized by three behavioral castes that assumed the roles of (1) brood care and colony maintenance, (2) social interactions and (3) foraging. Analysis of head widths indicated that workers were monomorphic but that there was an apparent size bias to division of labor with foragers being larger than brood care workers, and brood care workers being larger than those engaged in social interactions. A similar study of three colonies of the closely related and biologically similar species, *L. ambiguus*, revealed a similar pattern of morphology and division of labor but no correlation between size and behavior (Herbers, 1983).

In my study, I tested for size differences among foragers (defined here as any worker outside its nest in a laboratory testing arena) and nonforagers in 15 *L. longispinosus* colonies that had been obtained from various sites and maintained in the laboratory for various periods of time (Stuart, in press). In this study, foragers and nonforagers often differed significantly in size, but size and behavior were not consistently correlated across colonies: foragers were significantly larger than nonforagers in 6 colonies, significantly smaller in 6 colonies and not significantly different in 3 colonies. Clearly, this is not a pattern characteristic of size polyethism, but some explanation is required for the large number of significant differences that were observed.

In 5 colonies, workers had been marked, and I discovered that age cohorts in these colonies differed significantly in size and that there was a significant association between age and behavior: the older workers tended to be foragers and the younger workers tended to be nonforagers. This pattern is characteristic of age polyethism, and, as discussed above, age polyethism appears to be nearly universal among social insects. Moreover, size differences among age cohorts in monomorphic ant species have also frequently been observed and apparently results from differences in larval nutrition (Brian & Brian, 1951; Delage-Darchen, 1974; Gray, 1971; Markin & Dillier, 1971; Plateaux, 1970, 1971; Rissing, 1987).

Consequently, the combination of age polyethism and age-dependent size variability is also likely to be quite common and is likely to produce spurious correlations between size and behavior within colonies. Furthermore, if successive age cohorts within a colony can be either larger, smaller or not significantly different in size from one another, then the inconsistent correlations between size and behavior across colonies that I observed in my study would be obtained. It might also be expected that conspecific colonies that were collected at the same time and in the same place might tend to exhibit a similar pattern of age-dependent size variability (and a similar correlation between size and behavior) because of similar environmental influences on larval growth, whereas colonies collected at different times or in different places could be much more variable. Thus, the larger and more diverse sample of colonies that I used in my study revealed the inconsistency of the size-behavior correlation in *L. longispinosus,* while the more limited sample of colonies studied by Herbers and Cunningham (1983) did not. Also, the absence of information on the age of the workers in their study effectively prevented them from discovering the three-way association between age, behavior and size, evidence that might have led them to re-evaluate their conclusion regarding size polyethism.

It is noteworthy that all of the studies that purport to document size polyethism in monomorphic ants involve very few colonies and that none of them report the age of the workers involved (Breed & Harrison, 1988; Herbers & Cunningham, 1983; Hölldobler, 1981). Given the near universal importance of age in division of labor and the widespread influence of nutrition on larval growth and worker size (discussed above), none of these studies can be considered to provide sufficient evidence for size polyethism in monomorphic ants. Indeed, future studies that attempt to document additional influences on division of labor in social insects might be well advised to account for the possible confounding effects of age, or risk reporting similar spurious results.

Although current evidence is insufficient to demonstrate size polyethism in monomorphic ants, such size biases might well occur. Worker size could influence various physiological processes and has been shown to be correlated with differing metabolic rates and longevity within and among certain species (Bartholomew, Lighton & Feener, 1988; Calabi & Porter, 1989; Lighton, Bartholomew & Feener, 1987; MacKay, 1982) and could influence various physiological processes. These kinds of effects might be evident in monomorphic species, might impact on division of labor and could provide the substrate from which elaborate forms of worker polymorphism have evolved. Nonetheless, although expected, these potential associations have yet to be convincingly demonstrated.

Adaptive Demography and the Evolution of Castes
Wilson (1968) and Oster and Wilson (1978) developed a comprehensive theory of the evolution of castes within colonies. According to this theory, ergonomic efficiency is maximized if colonies maintain the optimal caste distribution function: the optimal proportions of various castes (both physical and temporal) based on a programmed schedule of birth, growth and death. Thus, this theory

predicts that evolution should give rise to an adaptive demography within colonies and that colony demographics should reflect ecological differences and have fitness consequences. Various lines of evidence support this theory, but, to date, explicit tests have been relatively rare and have produced mixed results (reviews in Beshers & Traniello, 1994; Herbers, 1980; Schmid-Hempel, 1992; Traniello, this volume). However, considering the complexity of the systems involved, the various ecological variables that might be important and the extensive variability and flexibility associated with division of labor (discussed below), it might be extremely difficult to design adequate tests of this theory for particular species. Moreover, given the degree of variability observed within and among species with respect to caste ratios and behavioral repertoire size (e.g., Johnston & Wilson, 1985; Wilson, 1984), it would be truly remarkable if colony demographics did not have fitness consequences and had not been shaped, at least to some extent, by natural selection. Considerable research remains to be done in this area. For recent discussions see Calabi (1988), Gordon (1989a), Schmid-Hempel (1992) and Traniello (this volume).

Behavioral Variability: Beyond Age and Size Polyethism

There are clear patterns of age and size polyethism in social insects, but there is also a good deal of variability with respect to division of labor, both within and among colonies, that is not explained by these two variables (Hölldobler & Wilson, 1990; Lenoir, 1987; Page & Robinson, 1991; Sudd, 1982). Workers pass through age castes at different rates with some showing apparent precocious behavioral development and others developing more slowly; the degree of task specialization within an age caste can be highly variable; and both age and size castes sometimes show considerable flexibility in response to changing colony needs (Bonavita-Cougourdon & Morel, 1988; Calabi, 1988; Calabi & Rosengaus, 1988; Calabi & Traniello, 1989; Gordon, 1989a, b; Hölldobler & Wilson, 1990; Lachaud & Fresneau, 1987; Lenoir, 1979, 1987; McDonald & Topoff, 1985; Page & Robinson, 1991; Porter & Tschinkel, 1985; Sorensen, Busch & Vinson, 1984). For example, experimental manipulations of colonies show that if large numbers of old or young workers are removed from colonies, then the remaining workers will exhibit either accelerated development or regression to fill in this gap. Indeed, even major workers that normally show a very limited behavioral repertoire will sometimes expand their repertoire if the colony is deprived of minor workers (e.g., Wilson, 1984). In an analogous fashion, some socially parasitic slave-making ants display expanded behavioral repertoires when deprived of their slaves (Stuart & Alloway, 1985). Harvester ant workers show a good deal of flexibility in task performance outside their nests and have been the subject of intensive study (Gordon, 1989a, b, 1991, 1992, this volume). This kind of variability in social insect colonies suggests that division of labor is a very dynamic process and that it is regulated in various ways and at various levels. Understanding this variability, and the factors and mechanisms that are responsible for it, should be a major goal in studies of division of labor.

Behavioral Genetics: Explaining Some of the Variability
Division of labor might well be expected to have a genetic component, and genetic components are a necessary precondition for the continuing evolution of this important aspect of social organization. Nonetheless, until recently, most discussions of division of labor totally ignored the possible existence of genetic components and described division of labor exclusively in developmental and environmental terms. To date, the existence and extent of genetic influences on division of labor have still received relatively little study in most social insects but have been the subject of considerable recent research in the honeybee. Page and Robinson and their associates (Page & Robinson, 1991; Page, Robinson, Britton & Fondrk, 1991) have shown that different subfamilies of honeybees reared within the same colony (i.e., offspring of the same queen but of different males) often differ in their tendency to perform various tasks and in the rate with which they progress through successive age castes. Thus, genetic variability is one possible cause of the variability in task performance that is often observed within and among colonies of social insects (discussed above). Recent experiments on various ant species provide further evidence for genetic components to division of labor (Carlin, Reeve, & Cover, 1993; Snyder, 1992; Stuart & Page, 1991). This research is discussed below.

In collaboration with Page, I conducted a series of cofostering

experiments with the monomorphic ant, *Leptothorax rudis*, to test
for a genetic component to division of labor in this species (Stuart
& Page, 1991). We used an experimental design that would effectively eliminate the influence of worker age on our results since age
has been shown to be such an important factor in division of labor
(discussed above). Worker pupae were removed from 3 pairs of field-
collected parental nests, and allowed to eclose in the absence of older
ants. At 1 or 2 day intervals, as these pupae eclosed, the young workers were removed, marked to identify colony of origin and combined to form subcolonies composed of equal numbers of workers
from each parental colony in the pair. For each pair of nests, 7 or 8
subcolonies were formed, for a total of 23 subcolonies. Thus, each
subcolony was composed of nearly identically aged workers from 2
different parental nests. These subcolonies appeared to function as
normal, small, queenless colonies: the workers foraged for food, laid
eggs and reared larvae. After allowing these subcolonies to age for
about 2 months, we compared the proportion of workers from each
parental nest that we observed foraging (defined here as any worker
outside its nest in a laboratory testing arena) or not foraging. The
pooled subcolonies from each of the 3 pairs of parental colonies all
showed significant differences in the tendency of workers to forage
depending on colony of origin. Since we controlled for possible age
effects, we interpret this experiment as providing strong evidence
for a genetic component to division of labor in this species. This
experiment did not control for possible differences in the larval rearing environments of the young workers (i.e., the different parental
nests and eclosion chambers) and this might be construed as a possible factor that might influence division of labor. However, this
potential difference has never been shown to have any impact on
division of labor for any social insect species.

Snyder (1992) provided further evidence for genetic components to division of labor in her study of the polygynous ant, *Formica
argentea*. Workers in 5 field-collected polygynous nests were individually marked, their behaviors were observed and they were later
assigned to familial groups based on an electrophoretic analysis. For
the six behaviors that were analyzed, there was a significant association between familial group and behavior for between one and three
behaviors for each colony. Overall, 9 of the 30 comparisons were

significantly different at the 0.05 (n = 6) or 0.01 (n = 3) level of significance based on a series of contingency table analyses and chi-square tests. Unfortunately, no adjustment of the experiment-wise error rate was made to compensate for the large number of comparisons being made (e.g., see Rice, 1989). Moreover, by using naturally constituted polygynous colonies, the ages of the workers in this study remain unknown, and possible heterogeneity in age distributions among familial groups within colonies is a possible alternative explanation for the results. The author counters a possible age explanation by noting that workers eclose during a brief 2–3 week period each year and that she was unable to detect any familial differences in eclosion pattern within one season in 2 colonies or any heterogeneity in the proportion of workers belonging to different familial groups across years in one colony. Nonetheless, these latter data pertain to very few colonies and do not apply directly to the workers that were involved in her study. Thus, it remains unclear to what extent age differences might have contributed to or confounded these results. In honeybees, age differences of as little as one day have been shown to impact on division of labor (Page et al., 1991), and the same might be true of ants.

Carlin, Reeve and Cover (1993) presented further evidence for genetic components to division of labor from a study of the polygynous ant, *Camponotus planatus*. Six colonies were collected and partitioned into monogynous subcolonies so that workers of known maternity could be reared. As new workers matured from eggs laid in these subcolonies, they were marked and combined to form experimental polygynous colonies corresponding to those that were originally collected. Observations revealed a strong tendency for behavioral specialization according to matriline. Indeed, matriline appeared to be a more important predictor of behavior than age in these colonies. However, since workers belonging to different matrilines were reared from egg to adult in separate subcolonies, it is possible that workers acquired some kind of predisposition to particular tasks as larvae or as newly eclosed adults that biased their behavior when combined in experimental colonies. Nonetheless, as previously stated, these potential differences in rearing conditions have never been shown to have any impact on division of labor.

Genetic influences on division of labor appear to be best interpreted as workers varying in their threshold for response to various stimuli, with the setting of the threshold having a genetic component (Page & Robinson, 1991). When a certain contingency arises, the most sensitive workers respond first, and, if the contingency is met, the least sensitive workers never respond because their threshold for response has not been reached. In a colony consisting of multiple families or subfamilies, there are probably different distributions of response thresholds among the individuals of each group. Under low stimulus conditions only the most sensitive workers in the most sensitive group respond. However, under a high stimulus situation, workers in all groups may respond. Developmental and environmental factors are also thought to impact on the setting of these response thresholds.

Further research is necessary to determine the existence of genetic influences on division of labor for other social insects and to evaluate the extent to which genetic variability can explain interindividual and intercolonial differences in division of labor. Comparative studies of genetic aspects of division of labor could be especially interesting in species that belong to phylogenetic groups where sociality evolved independently from honeybees, that exist within natural populations and under diverse ecological circumstances and that have different degrees of genetic relatedness in their colonies.

Hormonal Regulation: The Underlying Mechanism

There is now good evidence that juvenile hormone plays a role in the regulation of age polyethism in honeybees (Page & Robinson, 1991; Huang & Robinson, 1992). Low juvenile hormone titer is associated with behavior in the nest such as brood care, and high titers are associated with foraging. Artificial treatment with the juvenile hormone analogue methoprene can induce premature foraging in a dose-dependent manner. Robinson (1992) has proposed a model for the role of juvenile hormone in regulating honeybee age polyethism. In the model, juvenile hormone titer is influenced genetically (through the pattern of development) and by environmental conditions (especially colony conditions), and it subsequently influences the response threshold to certain task-associated stimuli, thereby modulating the response to those stimuli. Recent research has shown that interactions among honeybee workers belonging to different age cohorts can impact on juvenile hormone levels and can influence division of labor in predictable ways (Huang & Robinson, 1992). A similar role for juvenile hormone in the ontogeny of foraging behavior has been demonstrated in certain wasps (O'Donnell & Jeanne, 1993) and probably occurs in other social insects.

Self-Organization: A Substrate for Social Evolution

Page and Mitchell (1990) have recently suggested that certain features of insect societies that have previously been explained as direct products of natural selection might best be accounted for as expected outcomes of self-organization. Self-organization in complex systems has been the subject of recent work by Kauffman (1984, 1993), and these authors follow his approach by analyzing a social insect colony as a network of connected binary elements with boolean switching functions. Based on this analysis, some of the features of social insect colonies that might be attributed to self-organization are (1) homeostasis in which negative feedback reduces stimuli to near equilibrium as determined by the mean of certain threshold distributions, (2) mass action responses in which all elements turn on then off when simultaneous sampling is employed, (3) plasticity and resiliency in response to increases and decreases in external stimuli and (4) division of labor and specialization as demonstrated by steady state attractors where some individuals are frozen on and others are frozen off.

This analysis suggests that the minimal requirements for complex social organization might be only mutual tolerance and variability among individuals for thresholds for response to certain stimuli. Individuals would certainly be expected to have different response thresholds because of different learning experiences, different physiological states and different genotypes (Hölldobler & Wilson, 1990; Lenoir, 1987; Page & Robinson, 1991; Sudd, 1982); and various scenarios might lead to associations among individuals and result in such associations being perpetuated through natural selection (for discussions of the evolution of sociality see Hamilton, 1964; Brockmann, 1984). This analysis suggests that once such individuals associate, various fundamental features of social organization including division of labor might be expected to emerge as a direct result of this association without the necessity of any further intervention of natural selection. Subsequently, natural selection, operating at various levels, could act on any genetic components that contribute to this social organization and do considerable fine-tuning.

As stated, this concept of self-organization does not exclude natural selection from the origin, maintenance and elaboration of insect societies. Rather, it provides some indication of the kind of elementary substrate that might be expected to exist during the initial formation of insect societies and upon which natural selection might act. Page and Mitchell go on to suggest that the ordered sequences of behavior associated with egg laying in many solitary bees and wasps might be connected to the kinds of changes that we now observe played out over the lifetime of a worker honeybee, and that such changes might form the foundation for age polyethism.

Page and Mitchell provide what might be an important clarification of the relationship between self-organization and adaptation in insect societies. Self-organization also provides one explanation for the remarkable similarities that apparently occur among groups of apparently independently evolved social insects. Historically, many authors have complained about uncritical selectionist explanations of complex traits (Gould, 1978; Gould & Lewontin, 1979). The kind of model presented here may assist us in distinguishing between those traits that are and are not the products of natural selection.

Foraging for Work: Does Age Polyethism Require an Inherent Age Bias?

Tofts and Franks (1992) have proposed a model for division of labor in which age polyethism is viewed as an emergent property rather than as a fundamental organizational basis for division of labor. This model, dubbed the "foraging-for-work" model, proposes that the often systematic age-associated progression in task performance that is referred to as age polyethism, arises from the combined effects of (1) the structural organization of the nest, (2) random movements of workers among tasks in adjacent task zones and (3) the tendency of workers to continue to perform particular tasks as long as they continue to find work of that type that needs to be done. This model is extremely dynamic, permits a good deal of flexibility in the ongoing division of labor within colonies and can incorporate many recent observations regarding the process of division of labor, e.g., task specialization, genetic effects, juvenile hormone effects. However, the ultimate question raised by this research appears to be whether there is an inherent age bias to division of labor in social insect colonies. Does the apparent age effect constitute causation or mere correlation? Is there an age-related developmental process that biases division of labor and typically gives rise to what we call temporal polyethism?

These questions are well worth asking, and the answer could differ among species and among different social insect groups. However, the Tofts and Franks model does not present a clear alternative to concepts involving an inherent age bias in division of labor because it incorporates an age bias as one of its underlying assumptions. It assumes that workers are always born into the first task that is located on the brood pile, and that the young, newly eclosed workers passively displace older experienced workers from this task (Tofts & Franks, 1992, p. 347). Given this inherent age bias, it might not be too surprising that at least a weak form of age polyethism emerges from the model. Perhaps for species that exhibit only a very weak form of age polyethism, such a minor bias might be all that is involved.

Recently, Sendova-Franks and Franks (1993) and Tofts (1993) have provided further empirical and theoretical underpinnings respectively for this model, but they do not provide a convincing experimental test for the existence of an inherent age bias. The experimental study by Sendova-Franks and Franks (1993) is an intensive study of a single laboratory colony of *Leptothorax unifasciatus*, and, while some interesting results were obtained, they were not always consistent with the model. For example, some callows immediately appeared to become foragers, and it was suggested that the older, more experienced workers on the brood pile had displaced them to this activity (pp. 75, 94). Indeed, this result would appear to involve even less of an age bias than that incorporated into the current model (see above). However, it is unclear whether this result is a general pattern for this species or merely some aberration within this particular colony. Other *Leptothorax* species appear to exhibit much stronger forms of age polyethism (e.g., *L. longispinosus*, Stuart, in press), and it might be best not to generalize too quickly from studies of single laboratory colonies.

Robinson, Page and Huang (1994) have responded to the foraging-for-work model and summarized the various lines of evidence for an inherent and relatively strong developmental and age-related bias in division of labor in the honeybee. They argue that the foraging-for-work model might be consistent with the behavior of some species of social insects that exhibit only a weak form of age polyethism but that it lacks generality because it cannot account for the strong but flexible temporal polyethism observed in other species such as the honeybee. Thus, at present, it is unclear whether or to what extent the foraging-for-work model might appropriately be applied to division of labor in social insects.

Summary

Division of labor among workers in social insect colonies is a fundamental aspect of insect sociality and can be correlated with age size, or genetic differences among workers. However, division of labor is not rigidly determined. Rather, it is a very dynamic phenomenon and incorporates a good deal of variability and flexibility with respect to changing colony needs. Research directed toward understanding the various factors and mechanisms that influence division of labor continue to produce new and exciting results and to generate new hypotheses. Recent evidence for genetic effects in the division of labor of honeybees and various ant species provides an important explanation for some of the variability that can be observed within and among colonies. Genetic effects must now be considered along with developmental and environmental effects as one of the major classes of influences on division of labor in social insects.

The concept of self-organization in complex systems is also providing an important new perspective for explaining some aspects of division of labor. Indeed, the mere association of individuals with variable thresholds for response to various stimuli is likely to have produced some form of division of labor in even the most rudimentary of insect societies. Given that there were genetic components involved in the setting of these thresholds, then natural selection could act to further modify the dynamics of division of labor. Social insects are a large and diverse group, and studies of division of labor remain at a relatively early stage. As further species are studied, and as new hypotheses are refined and tested, many more discoveries are likely to be made.

References

Bartholomew, G. A., Lighton, J. R. B., & Feener, D. H. Jr. (1988). Energetics of trail-running, load carriage, and emigration in the column-raiding army ant *Eciton hamatum*. *Physiological Zoology, 61,* 57–68.

Beshers, S. N., & Traniello, J. F. A. (1994). The adaptiveness of worker demography in the attine ant *Trachymyrmex septentrionalis*. *Ecology, 75,* 763–775.

Bonavita-Cougourdon, A., & Morel, L. (1988). Interindividual variability and idiosyncrasy in social behaviors in the ant *Camponotus vagus* Scop. *Ethology, 77,* 58–66.

Breed, M. D., & Harrison, J. M. (1988). Worker size, ovary development and division of labor in the giant tropical ant, *Paraponera clavata* (Hymenoptera: Formicidae). *Journal of the Kansas Entomological Society, 61,* 285–291.

Brian, M. V., & Brian, A. D. (1951). Insolation and ant population in the west of Scotland. *Transactions of the Royal Entomological Society of London, 102,* 303–330.

Brockmann, H. J. (1984). The evolution of social behaviour in insects. In J. R. Krebs & N. B. Davies (eds.), *Behavioural ecology* (pp. 340–361). Oxford: Blackwell.

Calabi, P. (1988). Behavioral flexibility in Hymenoptera: A re-examination of the concept of caste. In J. C. Trager (ed.), *Advances in myrmecology* (pp. 237–258). Leiden: E. J. Brill.

Calabi, P., & Porter, S. D. (1989). Worker longevity in the fire ant *Solenopsis invicta*: Ergonomic considerations of correlations between temperature, size and metabolic rates. *Journal of Insect Physiology, 35,* 643–649.

Calabi, P., & Rosengaus, R. (1988). Interindividual differences based on behavior transition probabilities in the ant *Camponotus sericeiventris*. In R. L. Jeanne (ed.), *Interindividual behavioral variability in social insects* (pp. 61–89). Boulder, Colo.: Westview Press.

Calabi, P., & Traniello, J. F. A. (1989). Behavioral flexibility in age castes of the ant, *Pheidole dentata. Journal of Insect Behavior, 2,* 663–677.

Carlin, N. F., Reeve, H. K., & Cover, S. P. (1993). Kin discrimination and division of labor among matrilines in the polygynous carpenter ant, *Camponotus planatus*. In L. Keller (ed.), *Queen number and sociality in insects* (pp. 362–401). Oxford: Oxford University Press.

Delage-Darchen, B. (1974). Polymorphismus in der Ameisengattung Messor und ein Vergleich mit *Pheidole*. In G. H. Schmidt (ed.), *Sozialpolymorhismus bei Insekten,* pp. 590–603. Stuttgart: Wissenschaftliche.

Deneubourg, J. L., Goss, S., Pasteels, J. M., Fresneau, D., & Lachaud, J. P. (1987). Self-organizing mechanisms in ant societies (II): Learning in foraging and the division of labor. In J. L. Deneubourg & J. M. Pasteels (eds.), *From individual to collective behaviour in social insects,* Experimentia Supplementum, Vol. 54 (pp. 177–196). Basel: Birkhauser Verlag.

Gordon, D. M. (1989a). Caste and change in social insects. In P. H. Harvey & L. Partridge (eds.), *Oxford surveys in evolutionary biology,* Vol. 6 (pp. 55–71). Oxford: Oxford University Press.

Gordon, D. M. (1989b). Dynamics of task switching in harvester ants. *Animal Behavior, 38,* 194–204.

Gordon, D. M. (1991). Behavioral flexibility and the foraging ecology of seed-eating ants. *American Naturalist, 138,* 379–411.

Gordon, D. M., Goodwin, B. C., & Trainor, L. E. H. (1992). A parallel distributed model of the behavior of ant colonies. *Journal of Theoretical Biology, 156,* 293–307.

Gould, S. J. (1978). Sociobiology: The art of storytelling. *New Scientist, 80,* 530–533.

Gould, S. J., & Lewontin, R. C. (1979). The spandrels of San Marco and the Panglossian paradigm: A critique of the adaptionist program. *Proceedings of the Royal Society of London, 205,* 581–598.

Gray, B. (1971). A morphometric study of the ant species, *Myrmica dispar. Insectes Sociaux, 18,* 95–110.

Hamilton, W. D. (1964). The genetical evolution of social behavior. *Journal of Theoretical Biology, 7,* 1–52.

Herbers, J. M. (1980). On caste ratios in ant colonies: Population responses to changing environments. *Evolution, 34,* 575–585.

Herbers, J. M. (1983). Social organization in *Leptothorax* ants: Within and between species patterns. *Psyche, 90,* 361–386.

Herbers, J. M., & Cunningham, M. (1983). Social organization in *Leptothorax longispinosus* Mayr. *Animal Behaviour, 31,* 759–771.

Hölldobler, B. (1981). Foraging and spatiotemporal territories in the honey ant, *Myrmecocystis mimicus* Wheeler (Hymenoptera: Formicidae). *Behavioral Ecology and Sociobiology, 9,* 301–314.

Hölldobler, B., & Wilson, E. O. (1990). *The ants.* Cambridge, Mass.: Belknap Press of Harvard University Press.

Huang, Z. -Y., & Robinson, G. E. (1992). Honeybee colony integration: Worker-worker interactions mediate hormonally regulated plasticity in division of labor. *Proceedings of the National Academy of Science U.S.A., 89,* 11726–11729.

Jeanne, R. L. (1986a). The organization of work in *Polybia occidentalis*: Costs and benefits of specialisation in a social wasp. *Behavioral Ecology and Sociobiology, 19,* 333–341.

Jeanne, R. L. (1986b). The evolution of the organization of work in social insects. *Monitore Zoologico Italiano, 20,* 119–133.

Johnston, A. B., & Wilson, E. O. (1985). Correlates of variation in the major/minor ratio of the ant, *Pheidole dentata* (Hymenoptera: Formicidae). *Annals of the Entomological Society of America, 78,* 8–11.

Kauffman, S. A. (1984). Emergent properties in random complex automata. *Physica, 10D,* 145–156.

Kauffman, S. A. (1993). *The origins of order: Self organization and selection in evolution.* New York: Oxford University Press.

Lachaud, J. P., & Fresneau, D. (1987). Social regulation in ponerine ants. In J. L. Deneubourg & J. M. Pasteels (eds.), *From individual to collective behaviour in social insects,* Experimentia Supplementum, Vol. 54 (pp. 197–217). Basel: Birkhauser Verlag.

Lenoir, A. (1979). Feeding behavior in young societies of the ant *Tapinoma erraticum* L.: Trophalaxis and polyethism. *Insectes Sociaux, 26,* 19–37.

Lenoir, A. (1987). Factors determining polyethism in social insects. In J. L. Deneubourg & J. M. Pasteels (eds.), *From individual to collective behaviour in social insects,* Experimentia Supplementum, Vol. 54 (pp. 219–241). Basel: Birkhauser Verlag.

Lighton, J. R. B., Bartholomew, G. A., & Feener, D. H. Jr. (1987). Energetics of locomotion and load carriage and a model of the energy cost of foraging in the leaf-cutting ant *Atta colombica* Guer. *Physiological Zoology, 60,* 524–537.

MacKay, W. P. (1982). An altitudinal comparison of oxygen consumption rates in three species of *Pogonomyrmex* harvester ants (Hymenoptera: Formicidae). *Physiological Zoology, 55,* 367–377.

Markin, G. P., & Dillier, J. H. (1971). The seasonal life cycle of the imported fire ant, *Solenopsis saevissima richteri,* on the Gulf Coast of Mississippi. *Annals of the Entomological Society of America, 64,* 562–565.

McDonald, P., & Topoff, H. (1985). The social regulation of behavioral development in the ant *Novomessor albisetosus. Journal of Comparative Psychology, 99,* 3–14.

Moffett, M. W. (1988). Foraging dynamics of the group-hunting myrmicine ant, *Pheidologeton diversus. Journal of Insect Behavior, 1,* 309–331.

O'Donnell, S., & Jeanne, R. L. (1993). Methoprene accelerates age polyethism in workers of a social wasp (*Polybia occidentalis*). *Physiological Entomology, 18,* 189–194.

Oster, G. F., & Wilson, E. O. (1978). *Caste and ecology in the social insects.* Princeton, N.J.: Princeton University Press.

Page, R. E., & Mitchell, S. D. (1990). Self organization and adaptation in insect societies. In A. Fine, M. Forbes & L. Wessels (eds.) *Philosophy of Science Association 1990,* Vol. 2, (pp. 289–298).

Page, R. E., & Robinson, G. E. (1991). The genetics of division of labor in honey bee colonies. *Advances in Insect Physiology, 23,* 117–169.

Page, R. E., Robinson, G. E., Britton, D. S., & Fondrk, M. K. (1991). Genotypic variability for rates of behavioral development in worker honey bees (*Apis mellifera* L.). *Behavioral Ecology, 3,* 173–180.

Plateaux, L. (1970). Sur le polymophisme social de la fourmi *Leptothorax nylanderi* (Förster). I. Morphologie et biologie comparée des caste. *Annales des Sciences Naturelles, ser. 12, 12,* 373–478.

Plateaux, L. (1971). Sur le polymophisme social de la fourmi *Leptothorax nylanderi* (Förster). II. Activité des ouvrières et déterminisme des caste. *Annales des Sciences Naturelles, ser. 12, 13,* 1–90.

Porter, S. D., & Tschinkel, W. R. (1985). Fire ant polymorphism: The ergonomics of brood production. *Behavioral Ecology and Sociobiology, 16,* 323–336.

Rice, W. R. (1989). Analyzing tables of statistical tests. *Evolution, 43,* 223–225.

Rissing, S. W. (1987). Annual cycles in worker body size of the seed harvester ant *Veromessor pergandei* (Hymenoptera: Formicidae). *Behavioral Ecology and Sociobiology, 20,* 117–124.

Robinson, G. E. (1992). Regulation of division of labor in insect societies. *Annual Review of Entomology, 37,* 637–665.

Robinson, G. E., Page, R. E. Jr., & Huang, Z. -Y. (1994). Temporal polyethism in social insects is a developmental process. *Animal Behaviour, 48,* 467–469.

Schmid-Hempel, P. (1992). Worker castes and adaptive demography. *Journal of Evolutionary Biology, 5,* 1–12.

Sendova-Franks, A., & Franks, N. R. (1993). Task allocation in ant colonies within variable environments (A study of temporal polyethism: Experimental). *Bulletin of Mathematical Biology, 55,* 75–96.

Snyder, L. E. (1992). The genetics of social behavior in a polygynous ant. *Naturwissenschaften, 79,* 525–527.

Sorensen, A. A., Busch, T. M., & Vinson, S. B. (1984). Behavioral flexibility of temporal subcastes in the fire ant *Solenopsis invicta* in response to food. *Psyche, 91,* 319–331.

Stuart, R. J. (In press). Age, size, and division of labor in the monomorphic ant, *Leptothorax longispinosus. Behavioral Ecology and Sociobiology.*

Stuart, R. J., & Alloway, T. M. (1985). Behavioural evolution and domestic degeneration in obligatory slave-making ants (Hymenoptera: Formicidae: Leptothoracini). *Animal Behaviour, 33,* 1080–1088.

Stuart, R. J., & Page, R. E. Jr. (1991). Genetic component to division of labor among workers of a leptothoracine ant. *Naturwissenschaften, 78,* 375–377.

Sudd, J. H. (1982). Ants: Foraging, nesting, brood behavior, and polyethism. In H. R. Hermann (ed.) *Social insects,* Vol. 4 (pp. 107–155). New York: Academic Press.

Tofts, C. (1993). Algorithms for task allocation in ants (A study of temporal polyethism: Theory). *Bulletin of Mathematical Biology, 55,* 891–918.

Tofts, C., & Franks, N. R. (1992). Doing the right thing: Ants, honeybees and naked mole-rats. *Trends in Ecology and Evolution, 7,* 346–349.

Traniello, J. F. A. (1978). Caste in a primitive ant: Absence of age polyethism in *Amblyopone. Science, 202,* 770–772.

Tschinkel, W. R. (1988). Colony growth and the ontogeny of worker polymorphism in the fire ant, *Solenopsis invicta. Behavioral Ecology and Sociobiology, 22,* 103–115.

Wheeler, D. E. (1991). The developmental basis of worker caste polymorphism in ants. *American Naturalist, 138*, 1218–1238.

Wilson, E. O. (1968). The ergonomics of caste in social insects. *American Naturalist, 102*, 41–66.

Wilson, E. O. (1983). Caste and division of labor in leaf-cutter ants (Hymenoptera: Formicidae: *Atta*). IV. Colony ontogeny of *A. cephalotes. Behavioral Ecology and Sociobiology, 14*, 55–60.

Wilson, E. O. (1984). The relation between caste ratios and division of labor in the ant genus *Pheidole* (Hymenoptera: Formicidae). *Behavioral Ecology and Sociobiology, 16*, 89–98.

Wilson, E. O. (1985). The sociogenesis of insect colonies. *Science, 228*, 1489–1495.

Ecology, Colony Demography, and Social Organization in Ants

James F. A. Traniello

Dedicated to the memory of Robert Elliot Silberglied

One paradox in animal social biology is that groups composed of individuals of relatively high biological complexity tend to have a relatively low degree of cohesiveness and integrated action (Wilson, 1975a). Although individuals in vertebrate societies may forage in groups, communicate information about the presence of a predator and actively defend vulnerable group members, there appears to be little specialization in the behavioral repertoires of individuals. In the vast majority of vertebrate societies there are no castes, or groups of age- or size-correlated specialists in reproduction, offspring care, defense and foraging. Indeed, in vertebrate groups individuals attempt to maximize their own reproductive success within a social framework rather than maximize the fitness of the group as a whole. Kin-selected cooperation or reciprocally altruistic relationships among group members may add an additional tier of interaction and level of complexity. Socioecological studies have shown that vertebrate group structure, measured in terms of group size, composition and mating systems, is adaptive (reviewed in Krebs & Davies, 1992). Nevertheless, the organization of a vertebrate social group is still largely a consequence of the selfish reproductive interests of its members. Natural selection has not acted to favor an adaptive structure of the group as a whole.

In the social insects the specialization of individuals is common, and caste and division of labor are the hallmarks of sociality in ants, bees and wasps, although there are differences among groups in the breadth of individual repertoires. As social organization in vertebrates reflects adaptation to the environment, the organization of insect societies, which includes labor specialization, also appears to have ecological correlates. Wilson (1968) and Oster and Wilson (1978) established a comprehensive ergonomic theory of the origin and significance of patterns and processes of division of labor in the social insects to attempt to understand the design of a colony, which is the unit of selection in the social insects.

Eusociality created groups in which individuals largely subordinated their own reproduction to the interests and requirements of the colony as a whole; kin selection provided an advantage in inclusive fitness to do so (Hamilton, 1964). Eusociality in the ants and termites involves significantly more than castes that are nonreproductive through evolutionary design. Workers often exhibit extreme and unusual morphological modifications that are associated with colony defense. Massive heads bearing crushing jaws; elongated, scythe-shaped mandibles; or nozzle-like extensions of the head capsule to discharge defensive chemicals are common in many species. Caste theory predicts that the proportions of such specialists in a colony should reflect ecology; therefore the demography of a colony is itself adaptive (Beshers & Traniello, 1994; Calabi & Traniello, 1989a; Herbers, 1980; Oster & Wilson, 1978; Schmid-Hempel, 1992). Recent research has suggested that division of labor in ants may have a genetic component, thus lending support to the hypothesis that selection can indeed act on phenotypically and genetically diverse colonies to favor adaptive patterns of task performance (Stuart, this volume; Stuart & Page 1991).

The pervasive influence of caste theory on the study of division of labor in social insects in general, the many demonstrations of the ergonomic significance and ecological correlates of worker size variation (reviewed in Beshers & Traniello, 1994; Herbers, 1980; Schmid-Hempel, 1992) and studies on the genetic and physiological underpinnings of task performance (Robinson & Page, 1989; reviewed in Robinson, 1992; Stuart & Page, 1991) have led to the conceptualization of an insect colony as a social system having attributes shaped by selection acting at the colony level and bearing a high degree of plasticity. Task assignment thus has a size, age, genotype and social basis. Although worker behavior may be correlated with age and size, it is evident that worker age and size frequency distributions (the caste distribution function) does not determine an individual's task performance. In fact the way labor is allocated at the colony level may have little relationship to the age distribution of workers (Calabi & Traniello, 1989a). Demonstrations of accelerated behavioral development and regression (McDonald & Topoff, 1985; Calabi & Traniello, 1989b) and studies of the organizational rules of group-level dynamics (Gordon, this volume) also support the existence of important epigenetic principles of labor allocation.

Our research has centered on analyzing task allocation in ants to understand the degree to which colony-level selection has shaped worker demographic distributions. This is one approach to understanding the organizational basis of collective action. We have also attempted to understand how the environment affects individual and group behavior to better describe the factors that shape a worker's tendency to perform or specialize on a given task. Thus our approach has been to analyse the community ecology of colonies as well as the social environment of individuals in those colonies to reveal the linkage between individual and group behavior. In this paper I review the results of recent research and discuss their significance to caste theory, the process of division of labor and the organization of ant societies.

Colony Demography and Foraging Strategy in Ants

The role played by caste in colony ergonomics should be evident in the food-collecting activities of workers. For the latter portion of their lives, workers generally perform tasks outside of the nest, and foraging behavior often predominates in an individual's repertoire. And because workers are sterile, they do not experience the competing demands on their time and energy budgets that would be associated with mating or other activities, so their behavioral repertoires contain only a few acts. Caste theory holds that foragers have been designed by natural selection to optimize energetic return (Oster & Wilson, 1978), and both individual and social aspects of foraging strategy should reflect the the ergonomic benefits and costs of worker food-harvesting behavior and colony-level patterns of investment in workers of varying size (Traniello, 1989). Specifically, worker polymorphism is considered to be an adaptation of the caste distribution function (CDF) to the resource distribution function. That is, the CDF is thought to be "matched" to the pattern of distribution of food items in the environment in size, space and time and may also be shaped by the competitive environment (Davidson, 1978; Oster & Wilson, 1978; reviewed in Traniello, 1989).

Traniello and Beshers (1991) examined the relationship between forager size variation and forage load size selection in the Florida harvester ant *Pogonomyrmex badius* by collecting workers and their forage loads and correlating load size variables with forager size variables. We found no overall significant association between worker size and load size, although some forager size variables (head width and mass) were significantly correlated with food item mass in one of the two colonies sampled. These significant correlations were obtained for insect prey but not seeds, so polymorphism in *P. badius* may be associated more with an expansion of the diet into omnivory than an expansion of seed sizes collected. More recent studies by Ferster and Traniello (1995) tested the hypothesis that load size selection is related to forager size and the distance from the nest at which the item is collected (Reyes Lopez, 1987). But again data do not consistently support the notion of enhanced energetic efficiency of foraging through the proliferation of variation in size in the worker caste. One reason why such an association may not occur is the relative low energetic cost of foraging in *Pogonomyrmex*: energy expenditure during a foraging trip is a small fraction of the average energetic yield of a forage load (Fewell, 1988).

Foraging habits may also influence the association between worker age and task performance. In most species of ants studied to date, foraging is a task that begins relatively late in a worker's life, typically following inside-nest activities such as brood care (Wilson, 1985). In the ponerine *Amblyopone pallipes*, workers prey on vermiform arthropods such as centipedes and feed developing larvae by placing them directly on the prey (Traniello, 1978). Predatory behavior and brood care are therefore associated in this species, and foraging behavior is independent of worker age.

The Adaptiveness of Colony Demography in *Pheidole dentata*

According to the theory of adaptive demography, the caste distribution function of a colony has evolved to maximize reproductive success. In this model of ergonomic optimization, the frequencies of size- or age-related specialists represent adaptations to environmental contingencies faced by a colony. The role of caste in foraging behavior, as discussed above, is one aspect of the adaptiveness of colony demography. Older and/or larger workers may be chiefly involved in foraging whereas newly eclosed and/or smaller workers tend brood. Other workers may be defensive specialists that guard nest entrances or respond to the presence of a predator or competitor by showing aggressive behavior. These workers have morphological and behavioral adaptations (usually phragmotic heads or enlarged mandibles and reduced behavioral repertoires) that suit their defensive role. Colonies in different environments may face different patterns of food availability, competition or rates of predation. Ecology is thought to have shaped the CDF, but the question remains: do castes occur in proportions that enhance colony fitness?

Pheidole dentata is an ant species that has a completely dimorphic worker caste; minor workers tend to most tasks in the colony while majors specialize on defense. Within the minor worker caste, individual ants also change from within-nest to outside-nest tasks as they age (Wilson, 1976). Because of the apparently strong association between size and defensive specialization and age and brood care and foraging behavior, *P. dentata* appeared to be an ideal model system to examine the correlation between the CDF of colonies and their local ecologies.

Calabi and Traniello (1989a) assessed variation in colony structure, as reflected in the CDF, by collecting whole, queenright colonies from decayed-wood nests at two sites in Florida. Censuses were taken to determine the proportions of major and minor workers (physical caste markers) as well as the proportions of minor workers in different age castes (temporal caste markers). Ecological factors considered *a priori* to be important determinants of caste structure were the intensity of predation and competition, and differences in food distribution patterns. These aspects of the community ecology of *P. dentata* colonies occurring in the two Florida habitats were measured. To estimate food availability, pitfall, Berlese and light traps were used to capture potential invertebrate prey. Competition between *P. dentata* and sympatric ant species was determined by measuring the rate of removal of carbohydrate and protein baits, as well as by observing the relative abundance of ants at baits in each habitat. Predation was assessed from estimates of the relative abundance of commonly myrmecophagous vertebrates and quantifying anuran gut contents.

In spite of significant differences in the invertebrate biomass, ant diversity and distribution, the intensity of competitive interactions between *P. dentata* and sympatric ants, and the diversity and abundance of predators between the two habitats in which *P. dentata* colonies were collected, there were no significant differences in the social structure of colonies as reflected in the proportions of physical and temporal castes. Surprisingly, the presence in one of the habitats of a major competing ant species, *Solenopsis invicta*, also had no effect on the proportions of majors in colonies collected in that habitat. This result is particularly revealing because *P. dentata* colonies respond to *S. invicta* by specifically recruiting majors to defend the colony against this important "enemy" (Wilson, 1975b). No aspect of colony demographics was correlated with colony fitness: physical and temporal caste ratios had no significant association with the number of alate reproductives produced by a colony (Calabi and Traniello, 1989a).

The Adaptiveness of Colony Demography in the Fungus-Gardening Ant *Trachymyrmex septentrionalis*

The second species in which we have examined the adaptiveness of colony demography is a fungus-gardening ant. *Trachymyrmex septentrionalis*, an attine ant that nests in the soil in open habitats such as meadows and woodlands, is an excellent model species to test the theory of adaptive demography or other theories of size/task performance because it is weakly polymorphic (0.75–1.20 mm in head width) and its geographic distribution extends along most of the Atlantic coast. Because this species occurs in a wide range of ecological conditions that are likely to influence worker demography, macrogeographic comparisons of colony structure could be made. Recently we completed a series of studies on the ecology of caste in *T. septentrionalis* (Beshers & Traniello, 1994). The basic protocol in these studies was similar to that used to examine ecology and caste in *P. dentata*. In this case whole, queenright colonies of different sizes, together with their fungus combs, were excavated from populations of *T. septentrionalis* in Long Island, New York, and Lake Placid, Florida. The two populations had significant differences in the portion of the year during which colonies are active and also differed ecologically in the species composition and abundance of sympatric ants.

Following the collection of colonies in the field, the size-frequency distribution of workers in each colony was determined by measuring head width. Comparisons of colony demographies were made from the mean and standard deviations of worker size, and the skewness and kurtosis of the size-frequency distributions, which describe differences from normality. Alate biomass was measured in colonies from each population to assess the association between demographic distributions and fitness and to determine if a given caste distribution function optimized fitness. In this case the optimal caste ratio was considered to be that which occured in the colony having the highest fitness.

Sociometric analyses yielded the following results:

1.   colonies from both Long Island and Florida populations had unimodal size-frequency distributions showing minor departures from normality;

2.   colonies from the Long Island population had significantly larger mean and standard deviations of worker size; these two size distribution descriptors were sufficient to discriminate colonies from the two populations;

3.   colonies from the Long Island population showed no change in demography as a function of colony size; in the Florida population the mean and standard deviation of worker size increased with colony size;

4.   alates were produced in all colonies, independent of size, in both populations; females, but not males, were significantly larger in Long Island colonies; although alate production increased with colony size, there was no significant association between colony size and alate production in mature colonies.

In contrast to our research on caste and ecology in *P. dentata, T. septentrionalis* demography had a significant affect on fitness in both the northern and southern populations; fitness declined with distance from the optimal demographic distribution, suggesting that the CDF is adaptively shaped by the environment.

The Ecology and Evolution of Caste in *P. dentata* and *T. septentrionalis*

Environmental Variation and Selection for Adaptive Caste Structure

Are there ecological correlates to the social organization of ant colonies? Our research produced seemingly conflicting results. In *P. dentata*, detailed comparisons of physical and temporal caste structures in colonies in two ecologically different habitats were unable to identify any environmental variable associated with caste, or any coupling of caste to fitness. In *T. septentrionalis*, the CDF differed in colonies from north and south temperate populations and affected fitness.

In *P. dentata*, the distance between the two habitats was not great (on the order of 1–2 km), so in spite of the ecological variation we described, it must be emphasized that the *P. dentata* study was carried out on a microgeographic scale. Although mating flights in *Pheidole* appear to be localized (Wilson, 1957, for *P. sitarches*; Traniello & Adams, unpublished observations on *P. pilifera*), we cannot rule out the possibility that there was some degree of dispersal of reproductives from each habitat and thus gene flow between habitats. Genetic similarity in colonies from the two populations could account for similarities in the caste distribution functions of the colonies censused.

It should also be noted that differences in the competitive environment, predation pressure and resource availability could in combination select for colonies with essentially the same CDF. That is, predation by different species of mammalian myrmecophages in the two habitats and the presence of important competing ant species could each select for a similar CDF, as both would favor a skewing of physical caste ratios toward soldiers. In a similar way, the effects of differences in resource distribution patterns and variable competitive environments could select for the same CDF. Establishing ecological differences allows us to *approximate* the nature of environmental differences; such correlations only suggest potential causal relationships between ecology and caste. Because environmental variation is composed of a number of factors that may have additive or conflicting selection effects, the relationship between ecology and caste may be obscure even if such variation is carefully estimated.

The macrogeographic comparison of demography in *T. septentrionalis* avoided some of the pitfalls of the microgeographic comparison of *P. dentata* colonies, even though there were no detailed estimates of ecological variation in the New York and Florida populations. The two populations were, nevertheless, easily distinguished by strong faunal and seasonal differences and were genetically isolated. The effects of ecology on worker size variation could thus be interpreted in a straightforward manner.

Worker Size- and Age-Frequency Distributions and Colony Fitness

Colony demography and fitness were related in *T. septentrionalis* but not in *P. dentata*, but does this mean that these two species offer conflicting results for caste theory? Complete dimorphism in the worker caste is an attribute of the social organization of the genus *Pheidole*, which is one of the three ant genera with the highest species diversity. As we have argued (Calabi & Traniello, 1989a), to understand the origin of a completely dimorphic worker caste in this genus, it is important to consider selection acting on an *evolutionary* time scale. Within this time frame, developmental pathways were "reprogrammed" (Wheeler, 1991) by selection acting to favor colonies with enhanced defenses against predators and/or competitors. Here the concepts of division of labor and caste seem intuitive; complete dimorphism affords a high degree of specialization within the worker caste. Variation in the CDF may occur on an *ecological* time scale; there appear to be constraints on the extent to which colonies may be capable of adaptively adjusting the CDF. For example, although *Solenopsis* is a significant threat to *P. dentata*, colonies of *P. dentata* do not alter caste proportions (i.e., invest more heavily in major workers) when they are continually assaulted with *Solenopsis* workers, simulating an important change in the local environment of laboratory colonies (Johnston & Wilson, 1985). In *Camponotus floridanus*, the perception of a competitive environment may cause changes in brood rearing (Calabi & Nonacs, 1994; Nonacs & Calabi, 1992); this result provides evidence of a mechanism that could be involved in adaptively adjusting caste ratios. There is not enough information to draw any major conclusion, but the existence of mechanisms to monitor the competitive environment or respond to perturbations by maintaining physical caste ratios (Porter & Tschinkel, 1985; Wilson, 1983a) together with a lack of adaptive adjustment of caste ratios suggest that division of labor may be organized by social processes that do not involve altering brood rearing to change the rate of production of a given caste.

The macrogeographic comparison of *T. septentrionalis* colony polymorphism is insightful because this species is characterized by weak size variation, which may represent a rudimentary condition in the evolution of worker polymorphism. Also, we are able to assess the adaptiveness of the CDF in a primitive species of fungus-growing ant, the group that includes *Atta*, a focal species in the study of the evolution of polymorphism (Wilson, 1980, 1983a, b). Colony size in *T. septentrionalis* was smaller in Long Island, and workers were larger and more variable in size than in Florida. Colonies in each population had distinct growth patterns as well; the demographic distributions were correlated with fitness.

But in *T. septentrionalis* this adaptive polymorphism appears to have little to do with the process of division of labor. All size classes perform virtually the entire repertoire of foraging, brood care, fungus rearing and colony maintenance tasks (Beshers & Traniello, in press). This is in sharp contrast to the strong task/caste associations found in *Atta*, in which minims care for the fungus gardens, intermediate-size workers cut and/or carry leaf fragments and large workers provide defense (Wilson, 1980). Also in *T. septentrionalis* a broader polymorphic range of workers does not appear to contribute significantly to colony survivorship or fitness. And within the genus *Trachymyrmex*, worker body size is highly variable in spite of the apparently high degree of similarity in colony requirements for brood and fungus care and nest maintenance.

Our study of caste polymorphism in *T. septentrionalis* provides an example of how polymorphism may be adaptive yet unrelated to the process of division of labor. Large worker body size should be favored over colony investment in a greater number of workers when abiotic factors such as a long period of thermal stress are of primary importance (Beshers & Traniello, 1994). *T. septentrionalis* colonies in the Long Island population face a longer period of inactivity (an eight-month diapause) at lower temperatures than do colonies from the Florida population. Larger workers live longer and have low rates of respiration (e.g., Calabi & Porter, 1989), so colonies with demographic distributions that include larger workers may have enhanced survivorship during the winter diapause. In this regard it is significant that female alate reproductives are larger in colonies in the Long Island population whereas male alate reproductives are not. Newly inseminated female reproductive rely upon their own energy reserves during the incipient stages of colony foundation, a period that encompasses stressful winter months. Males, in contrast, are only active during the mating flight and die soon after, so male size should be influenced only by the requirements for dispersal by flight and reproductive competition (the latter may or may not occur in this species).

Colonies of *T. septentrionalis* from the Florida population have larger colony size and smaller workers and produce a larger number of alates. This suite of traits seems consistent with the hypothesis that colonies in the Florida population are selected for rapid growth. Critical information on the nature of the competitive and predatory environments faced by *T. septentrionalis* colonies in Florida is, however, lacking.

Caste, Division of Labor and Behavioral Mechanisms of Task Allocation in Ants

Adaptation of the form and behavior of individuals to colony requirements varies on evolutionary and ecological time scales. Physical and age caste proportions change through evolutionary time as coarse-grained adaptations, but the CDF itself may not be able to show adaptive variability in the face of changing local ecology because environmental fluctuation may occur within a time frame that is too brief or unpredictable (Calabi & Traniello, 1989a). Extrinsic (ecological) perturbations may lead to internal (social) changes in colony organization. For example, Wilson (1983a) showed that colonies of *Atta sexdens* that were experimentally altered by removing certain series of polymorphic workers returned to the CDF that characterizes incipient colonies. Also, different species of *Pheidole* respond to the removal of minor workers by altering the behavioral repertoires of major workers (Wilson, 1984). Minor workers normally perform brood care tasks; when they are absent, major workers, which are defensive specialists, begin to groom and feed larvae.

The caste ratio at which major workers show this behavioral change appears to be species specific, suggesting that differing ecological circumstances have set the thresholds for a major's repertoire change.

Temporal caste ratios may be sensitive to environmental perturbations as well. An intraspecific or interspecific conflict at a food source or territorial boundary could alter an age caste ratio if older workers were disproportionately involved in foraging and territorial defense and experienced high risk of injury. Temporal caste ratios may also be sensitive to the queen's oviposition rate, which has an affect on the rate of recruitment of new workers into the labor force and influences the requirement for brood care (Traniello, Beshers & Lawlis, unpublished). Oviposition rate and the timing of the development of brood could have more pronounced effects on caste ratios and colony needs in small colonies, which do not have a large worker population that can meet changing demands. It may also be the case that small and large colonies experience the same type of demands and respond in similar ways on a proportional scale.

Caste Distribution Functions and Colony Labor Profiles

Our socioecological studies of caste suggested that a tight coupling of age and size to task performance was unlikely. To more precisely describe the association between the CDF and colony-wide behavior, we conducted a series of ethological observations on *P. dentata* colonies. Differences among workers in task performance patterns have been estimated using a number of indices of the behavior of castes and individuals (Wilson, 1976; Seeley, 1982; Kolmes, 1985). We used a modified version of Seeley's (1982) relative performance measure to study the contribution of individuals in different temporal and physical castes to the labor profile of a colony. A colony labor profile is essentially the summation of the contribution of individuals to the performance of a given task (Calabi et al., 1983; Calabi & Traniello, 1989a).

The labor profiles of *P. dentata* colonies suggest that the social organization of division of labor and task performance is more complex than a pattern generated by straightforward size/task and age/task associations. First, studies of the age/task association showed that there were no differences between colonies collected in the two habitats in the relative performance of either inside-nest or outside-nest tasks. And although spatial association has been considered to be an important factor in the evolution of age castes, there were no significant associations in space between young, old or any caste and immature and mature colony members. In fact, colonies having nearly identical age caste proportions could have very different labor profiles, and colonies having very different age caste proportions could have similar labor profiles (Calabi & Traniello, 1989a). Second, patterns of worker social development in *P. dentata* are known to be affected by the existing distribution of age castes in a colony (Calabi & Traniello, 1989b). As has been demonstrated in honeybees and other species of ants (McDonald & Topoff, 1985), workers in colonies with experimentally skewed age distributions either accelerate their behavioral development or revert to performing tasks typically scheduled earlier in their ontogenic sequence. These results accentuate the fluidity and dynamic nature of task performance patterns and suggest an uncoupling of patterns of division of labor from the worker age variation that is predicted to drive task allocation.

In *T. septentrionalis* size variation in the worker caste is different in New York and Florida colonies and is correlated with fitness but does not appear to bear any relationship to the process of division of labor, so again an uncoupling of the CDF and task performance is indicated. In *T. septentrionalis* the worker caste can be divided into five age classes based on cuticular pigmentation differences, and workers can be separated on the basis of size into five classes. Independent of size, workers have broad behavioral repertoires and perform virtually all tasks (Beshers & Traniello, submitted). Neither worker age nor size appears to be associated in important ways with colony-wide patterns of division of labor.

Thus in *P. dentata* age and physical caste ratios are not correlated with colony fitness and are often not associated with colony behavior profiles, and in *T. septentrionalis* the degree of worker polymorphism does have a fitness correlate but physical polymorphism and temporal polyethism seem only weakly related to division of labor. Caste theory is inconsistently supported. What then are the organizational principles of division of labor in ants?

Caste theory does indeed provide an understanding of the organizational basis of division of labor, emphasizing the role of growth and efficiency. The dynamics of colony growth and investment in different castes may vary with colony life history and size, and different species may show different patterns in response to different selective forces acting at different times. Recently, Tschinkel (1993) used data from several ant species (*Camponotus impressus, Pheidole desertorum, Pogonomyrmex badius,* and *Solenopsis invicta*) to show that the percentage of majors varies interspecifically, and for each species the rate of investment in majors depends upon colony size. For example, in *P. badius,* the rate of production of majors increases sharply at a colony size of approximately 1000 workers, whereas the production of majors in *Solenopsis invicta* begins in colonies of about 100 workers and changes less dramatically as colony size increases. In *C. impressus* the proportion of majors decreases with colony size (Walker & Stamps, 1986). Although the nature of the selective forces acting at different times on colonies of different species are elusive (with the exception of changes in demography accompanying the production of sexuals), these data do provide evidence of the evolutionary nature of caste. The fact that fitness/demography correlates have been irregular and perhaps unrelated to division of labor should not be disturbing but rather encourage additional studies because sample size are minute in comparison to the number of ant species or social insect species in general.

Understanding the organization of caste and colony growth is but one approach to understanding polyethism. A second approach emphasizes the behavior patterns of colonies without focusing on age- or size-related task performance. Gordon (1984a) showed that five sympatric species of desert seed harvester ants have species-specific behavioral rhythms, which are related to the ecology and colony sizes characteristic of each species. Task performance may be unrelated to worker size, as is apparently the case in exterior workers of *P. badius;* some activities are consistently performed by the same group of workers while others change tasks, presumably in response to changing colony needs (Gordon, 1984b). In *P. barbatus,* the daily temporal pattern of task performance by workers outside of the nest (the "daily round") somewhat resembles one or more states of equilibrium. The priorities of colony needs were revealed through a series of perturbations that affect the numbers of workers engaged in certain tasks (Gordon, 1986), illustrating the interdependence of activities in a colony. Nest maintenance and foraging, for example, have a reciprocal relationship; an increase in the demand for one activity is followed by a decrease in the performance of the other. In this study and others, colony needs are manipulated to examine the dynamics of task switching. It is more difficult to understand the extent and dynamics of task switching and the patterns of task assignment that occur in undisturbed colonies. In undisturbed colonies, how accurate are worker age and size as predictors of task performance and to what extent do local or global colony needs override any typical pattern of size- and age-correlated behavior? Is task assignment different in incipient than in mature colonies? If behavioral flexibility is more commonly involved in task performance than demographic determination, does each process of task allocation have the same ergonomic efficiency? Does behavioral flexibility have limitations that have fitness correlates?

We could easily generate more questions of this type and propose numerous mechanisms of task allocation. It is clear that in spite of the long history of caste theory, the number of species studied and the number of questions answered are few and the study of division of labor is still in its infancy. In particular we need to reconcile the approaches of caste theory with those of task allocation theory. It is also critically important to examine the process of division of labor in reference to various ecological time scales (see also Gordon, 1991). The ecology of daily and seasonal activities may select for behavioral adaptations that are different from those occurring on an evolutionary time scale. Intraspecific comparisons of the age- and size-related behavior of individuals of different castes on daily and seasonal time scales at different points in a colony's ontogeny in different environments should prove valuable.

Acknowledgments

I thank the Whitehall Foundation and the National Science Foundation for supporting my research program. Sam Beshers and Simon Robson provided helpful comments on the manuscript. I thank the late Bob Silberglied for sharing with me his encyclopedic knowledge of insects.

References

Beshers, S. N., & Traniello, J. F. A. (1994). The adaptiveness of worker demography in the weakly polymorphic ant *Trachymyrmex septentrionalis*. *Ecology, 75,* 763–775.

Beshers, S. N. & Traniello, J. F. A. (in press). Polyethism and the adaptiveness of worker size variation in the attine ant *Trachymyrmex septentrionalis*. *Journal of Insect Behavior, 9*, 61–84.

Calabi., P., & Nonacs, P. (1994). Changing colony growth rates in *Camponotus floridanus* as a behavioral response to conspecfic presence. *Journal of Insect Behavior, 7*, 17–27.

Calabi, P., & Porter, S. D. (1989). Worker longevity in the fire ant *Solenopsis invicta*: ergonomic considerations of correlations between temperature, size, and metabolic rates. *Journal of Insect Physiology, 35*, 643–650.

Calabi, P., & Traniello, J. F. A. (1989a). Social organization in the ant *Pheidole dentata*: physical and temporal castes lack ecological correlates. *Behavioral Ecology and Sociobiology, 24*, 69–78.

Calabi, P., & Traniello, J. F. A. (1989b). Behavioral flexibility in the age castes of the ant *Pheidole dentata*. *Journal of Insect Behavior, 2*, 663–677.

Calabi, P., Traniello, J. F. A., & Werner, M. H. (1983). Age polyethism: Its occurrence in the ant *Pheidole hortensis* and some general considerations. *Psyche, 90*, 395–412.

Davidson, D. W. (1978). Size variability in the worker caste of a social insect (*Veromessor pergandei*) as a function of the competitive environment. *American Naturalist, 112*, 523–532.

Ferster, B., & Traniello, J. F. A. (1995). Polymorphism and foraging behavior in *Pogonomyrmex badius* (Hymenoptera : Formicidae) worker size, foraging distance, and load size associations. *Environmental Entomology, 24*, 673–678.

Fewell, J. H. (1988). Energetic and time costs of foraging in harvester ants, *Pogonomyrmex occidentalis*. *Behavioral Ecology and Sociobiology, 22*, 401–408.

Gordon, D. M. (1984a). Species-specific patterns in the social activities of harvester ant colonies (*Pogonomyrmex*). *Insectes Sociaux, 31*, 74–86.

Gordon, D. M. (1984b). The persistence of role in exterior workers of the harvester ant, *Pogonomyrmex badius*. *Psyche, 91*, 251–265.

Gordon, D. M. (1986). The dynamics of the daily round of the harvester ant colony (*Pogonomyrmex barbatus*). *Animal Behaviour, 34*, 1402–1419.

Gordon, D. M. (1991). Behavioral flexibility and the foraging ecology of seed-eating ants. *American Naturalist, 138*, 379–411.

Hamilton, W. D. (1964). The genetical evolution of social behaviour, I, II. *Journal of Theoretical Biology, 71*, 1–52.

Herbers, J. M. (1980). On caste ratios in ant colonies: Population responses to changing environments. *Evolution, 34*, 575–585.

Johnston, A., & Wilson, E. O. (1985). Correlates of variation in the major/minor ratio in the ant *Pheidole dentata* (Hymenoptera: Formicidae). *Annals of the Entomological Society of America, 78*, 8–11.

Kolmes, S. A. (1985). An ergonomic study of *Apis mellifera* Hymenoptera: Apidae). *Journal of the Kansas Entomological Society, 58*, 413–421.

Krebs, J. R., & Davies, N. B. (1992). *An introduction to behavioural ecology*. Sunderland, Mass.: Sinauer.

McDonald, P., & Topoff, H. (1985). The social regulation of behavioral development in the ant *Novomessor albisetosus*. *Journal of Comparative Psychology, 99*, 3–14.

Nonacs, P., & Calabi, P. (1992). Competition and predation risk: Their perception alone affects ant colony growth. *Proceedings of the Royal Society of London, Series B, 249*, 95–99.

Oster, G. F., & Wilson, E. O. (1978). *Caste and ecology in the social insects*. Princeton, N.J.: Princeton University Press.

Porter, S. D., & Tschinkel, W. R. (1985). Adaptive value of nanitic workers in newly founded colonies of the imported fire ant *Solenopsis invicta*. *Annals of the Entomological Society of America, 79*, 723–726.

Reyes Lopez, J. L. (1987). Optimal foraging in seed-harvesting ants: Computer-aided simulation. *Ecology, 68*, 1630–1633.

Robinson, G. E. (1992). Regulation of division of labor in insect societies. *Annual Review of Entomology, 37*, 637–665.

Robinson, G. E., & Page, R. E. (1989). Genetic basis for division of labor in an insect society. In M. Breed, & R. E. Page (eds.), *Genetics of social evolution* (pp. 61–80). Boulder, Colo.: Westview Press.

Schmid-Hempel, P. (1992). Worker castes and adaptive demography. *Journal of Evolutionary Biology, 5,* 1–12.

Seeley, T. D. (1982). Adaptive significance of the age polyethism schedule in honeybee colonies. *Behavioral Ecology and Sociobiology, 11,* 287–293.

Stuart, R. J., & Page, R. E. (1991). Genetic component to division of labor among workers of a leptothoracine ant. *Naturwissenschaften 78,* 375–377.

Traniello, J. F. A. (1978). Caste in a primitive ant: absense of age polyethism in *Amblyopone*. *Science, 202,* 770–772.

Traniello, J. F. A. (1989). Foraging strategies of ants. *Annual Review of Entomology, 34,* 191–210.

Traniello, J. F. A., & Beshers, S. N. (1991). Polymorphism and size-pairing in the harvester ant *Pogonomyrmex badius*: a test of the ecological release hypothesis. *Insectes Sociaux, 38,* 121–127.

Tschinkel, W. R. (1993). Sociometry and sociogenesis of colonies of the fire ant *Solenopsis invicta* during one annual cycle. *Ecological Monographs, 63,* 425–457.

Walker, J., & Stamps, J. (1986). A test of optimal caste ratio theory using the ant *Camponotus (Colobopsis) impressus*. *Ecology, 67,* 1052–1062.

Wheeler, D. E. (1991). The developmental basis of worker polymorphism in ants. *American Naturalist, 138,* 1218–1238.

Wilson, E. O. (1957). The organization of a nuptial flight of the ant *Pheidole sitarches* Wheeler. *Psyche, 64,* 46–50.

Wilson, E. O. (1968). The ergonomics of caste in the social insects. *American Naturalist, 102,* 41–66.

Wilson, E. O. (1975a). Sociobiology: the new synthesis. Cambridge, Mass.: Harvard Univ. Press.

Wilson, E. O. (1975b). Enemy specification in the alarm-recruitment system of an ant. *Science, 190,* 798–800.

Wilson, E. O. (1976). Behavioral discretization and the number of castes in an ant species. *Behavioral Ecology and Sociobiology, 1,* 141–154.

Wilson, E. O. (1980). Caste and division of labor in leaf-cutter ants (Hymenoptera: Formicidae: *Atta*). I. The overall pattern in *A. sexdens. Behavioral Ecology and Sociobiology, 7,* 143–156.

Wilson, E. O. (1983a). Caste and division of labor in leaf-cutter ants (Hymenoptera: Formicidae: *Atta*). III. Ergonomic resiliency in foraging by *A. cephalotes. Behavioral Ecology and Sociobiology, 14,* 47–54.

Wilson, E. O. (1983b). Caste division of labor in leaf-cutter ants (Hymenoptera: Formicidae: *Atta*). IV. Colony ontogeny of *A. cephalotes. Behavioral Ecology and Sociobiology, 14,* 55–60.

Wilson, E. O. (1984). The relation between caste ratios and division of labor in the ant genus *Pheidole* (Hymenoptera: Formicidae). *Behavioral Ecology and Sociobiology, 16,* 89–98.

Wilson, E. O. (1985). The sociogenesis of social insect colonies. *Science, 228,* 1489–1495.

Section III

Social Parasitism in Ants

Adaptations for Social Parasitism in the Slave-Making Ant Genus *Polyergus*

Howard Topoff

To many people, there is something unsettling about a parasitic lifestyle, in which one organism habitually exploits another to secure its food, shelter or even its reproductive hormones. But for the comparative psychologist, parasites provide a unique opportunity to elucidate the evolution and development of social bonds between different species of organisms. Strictly speaking, parasitism is a form of symbiosis, a broad ecological concept that also includes mutually beneficial associations, such as between a human being and the intestinal microorganisms that aid its digestion. But parasitism is decidedly one-sided, with the parasite living at the expense of its host, and sometimes killing it. People are most familiar with physiological parasites, like viruses or fleas, which attach themselves to the skin or internal organs of larger organisms. But an even more fascinating interspecific relationship among animals is known as social parasitism. At first, "social parasitism" sounds like an oxymoron, because the term "social" denotes communication, cooperation and even altruism, all diametrically opposed to the patently selfish habits of parasites. The term is appropriate, however, because a social parasite's infiltration into the host's life is based on the same developmental and communicative processes that the parasite and host use for communicating with members of their own species.

Among vertebrates, the best known of these symbionts are avian brood parasites, such as cuckoos (Davies & Brooke, 1988). In these species, the female parasite lays an egg in the nest of another species and leaves it for the host to rear, typically at the expense of the host's young. Less known but more varied are dulotic (from the Greek word for servant) species of insects in the order Hymenoptera (comprising bees, wasps and ants). Consider, for example, the unusual behavior of parasitic ants belonging to the genus *Polyergus*, which have lost the ability to care for themselves. The workers do not forage for food, feed their brood or queen or even clean their own nest. To compensate for these deficits, *Polyergus* has become specialized at obtaining workers from the related genus *Formica* to do these chores for them. This is accomplished by a slave raid, in which several thousand *Polyergus* workers travel up to 150 m, penetrate a *Formica* nest, disperse the *Formica* queen and workers and capture the resident's pupal brood. Back at the *Polyergus* nest, some raided brood is consumed. But a portion of it is reared through development, and the emerging Formica workers then assume all responsibility for maintaining the permanent, mixed-species nest (Talbot, 1967; Topoff, LaMon, Goodloe & Goldstein, 1984). They forage for nectar and dead arthropods and regurgitate food to colony members of both species. They also remove wastes and excavate new chambers as the population increases. Their focal role in the colony is also revealed when the worker population becomes too large for the existing nest. It is scouts of *Formica* that locate a new nesting site, return to the mixed-species colony and recruit additional *Formica* nestmates. During a period that may last 7 days, the *Formica* slaves then carry to the new nest all the *Polyergus* eggs, larvae and pupae; every *Polyergus* adult; and even the *Polyergus* queen.

Of the approximately 8,000 species of ants in the world, *Polyergus* is but one of over 200 species that have evolved some degree of symbiotic relationship with other ants (Alloway, 1980; Buschinger, 1986; Wilson, 1975). The distribution of these species among the 11 living ant subfamilies is quite uneven, as most parasitic species are confined to either the *Myrmicinae* (about 33 genera) or the *Formicinae* (9 genera). At one end of the behavioral continuum are temporary parasites, species capable of caring for themselves and relying on a host species only during the early stages of colony founding. For example, a newly mated queen of the formicine ant *Lasius umbratus* enters a nest of its host *Lasius niger*, kills the resident queen and deposits her own eggs in the invaded nest. The host workers rear her offspring that, as adults, scavenge for their own food. Because the host queen is no longer present, the worker force of *L. niger* gradually diminishes through attrition, and the colony becomes a single-species society of *L. umbratus* (Wilson, 1971).

At the other extreme of social parasitism are the inquilines, species of ants that spend their entire life cycle in the nest of the host. In *Teleutomyrmex schneideri*, for example, the entire worker caste has been eliminated and so have the slave raids. The queens of *Teleutomyrmex* are about one-third the size of their host queen, *Tetramorium caespitum*, and capitalize on their diminutive bulk by riding on the back of their hosts. The males and new queens produced by the parasitic female copulate inside the host nest. The newly mated queens then locate other colonies of *Tetramorium* to parasitize, and the cycle of parasitism is repeated.

A cardinal rule in evolutionary biology is that parasitic organisms, be they bacteria, tapeworms or fleas, are derived from free-living ancestors, and there is no reason to doubt that this principle holds as well for slave-making ants. Aside from this generalization, the evolution of socially parasitic ants has resulted in the elaboration of several competing hypotheses, drawing upon research in areas ranging from population biology to behavioral development.

Any theory explaining the evolution of obligatory social parasites must consider at least four processes. The first is a proficiency for locating target nests of slave species and for conducting group raids that result in brood capture. Second is the establishment of interspecific communication channels, so that slave ants remain in the colonies of their captors and routinely work for them. Third is a capability by queens for nonindependent colony foundation. For queens of free-living ants, colony reproduction is straightforward. After mating, a fertile female excavates a chamber, lays a few eggs and nourishes her larvae with stored nutrients. When the first brood matures into adult ants, these workers undertake general colony maintenance. An obligatory parasitic queen, on the other hand, requires the help of slaves to rear even her first brood. Therefore, these queens also must invade a target colony, drive off (or even kill) the host queen, appropriate the pupal brood and become accepted by the slave species' workers. If successful, these resident workers will rear the broods of the parasitic queen, until her worker population is sufficiently large to supplement the slave force by staging raids on other host colonies. Finally, the fourth process that an evolutionary hypothesis must explain is the behavioral sequence by which parasitic species evolved from free-living ancestors.

The Role of Scouts in the Organization of Slave Raids

Ever since the pioneering studies on *Polyergus rufescens* by Huber (1810) and Emery (1908), on *P. lucidus* by Talbot (1967) and Harman (1968) and on *P. breviceps* by Wheeler (1916), it has been well known that slave-making raids are initiated by a small group of workers called scouts. These individuals locate target colonies of *Formica*, return to their colony of origin, recruit nestmates and lead the raiders back to the *Formica* nest.

Our more recent field studies of *P. breviceps* were conducted at the Southwestern Research Station of The American Museum of Natural History, located 5 km west of Portal, Arizona. At an altitude of 1646 m, the ground in this habitat is covered with bunch grass and contains extensive leaf litter from alligator juniper, Arizona oak and Chihuahua pine. During one 6-week study, 18 Polyergus scouts were followed on 12 different days (Topoff, Bodoni, Sherman & Goodloe, 1987). Of these 18 scouts, 5 were tracked only on their outbound trip to a target colony of *Formica gnava*. For 3 other scouts, we succeeded in following their return trip as well. On 5 occasions, our tracking was abruptly halted when the scout was seized and killed by a spider. The remaining 5 scouts disappeared beneath the leaf litter before reaching a target colony.

The outbound path of a scout can be divided into two stages. Phase one begins when an individual leaves the swarm of ants circling around the nest and runs in a relatively constant compass direction for 25–45 m. The scout moves continuously during this period, without stopping to search for *Formica* colonies. The second phase incorporates a qualitatively different behavioral pattern, with the scout making looping movements over all compass directions. More importantly, it is only during this looping phase that scouts frequently stopped and searched beneath rocks and patches of leaf litter.

Polyergus is similar to other obligatory parasites in that workers do not search for food. Nevertheless, the location of target nests by scouts can be thought of as indirect foraging, because at least some of the raided *Formica* brood is fed to the *Polyergus* workers and queen by their resident slaves. It is therefore not surprising that the searching pattern of *Polyergus* scouts is virtually identical to that described for other formicine ants, such as *Cataglyphis bicolor*, which forages alone for dead arthropods (Harkness & Maroudas, 1985). In this desert-dwelling species, foragers also move away from the nest in a linear path, followed by a more tortuous looping search. A *Cataglyphis* forager can continuously keep track of its own position relative to home by using a path integration process (also known as dead reckoning). Furthermore, the ant's path integration system is backed up by a piloting process that relies on familiar landmarks (Wehner & Srinivasan, 1981). In addition, the linear phase of the scout's route ensures minumum overlap among the sectors searched by all of the scouts on any given day. *Polyergus* is also similar to *Cataglyphis* (and other formicine ants) in its ability to orient to polarized light, and we recently demonstrated that scouts use visual orientation during their outbound and return run, and when leading the raid swarm back to the target nest (Topoff et al., 1984). Further support for this hypothesis stems from our observations that scouts may use three different paths when they: (1) leave the nest for scouting, (2) return to the nest after locating a target colony and (3) lead the raiding swarm back to the target *Formica* colony. A model based upon chemical cues would predict a single path for the outbound run, return trip and slave raid, respectively.

An additional procedure of marking scouts with fluorescent powder verified that a scout runs intermittently at the head of each slave-raid swarm. On 3 occasions, we placed a glass jar over the marked scout midway through a slave raid. Each time this was done, the entire swarm came to an abrupt halt. The workers fanned out in all directions and started circling in much the same way they do near the nest entrance prior to raid onset. After 1 minute we removed the jar. The released scout promptly started running excitedly among the dispersed ants, then moved into the lead with the slave raiders (in a newly organized swarm) following closely behind.

The interaction between a marked scout and the raiders was also observed during an unsuccessful raid, when the column split around a large mesquite bush. The scout, followed by approximately two-thirds of the raiders, advanced for 3 m around the north end of the bush. There was no traffic between the two groups until the ants met and recombined at the far end of the mesquite. Thus the ants are capable of maintaining a compass bearing (at least for a short distance) without a scout. The swarm proceeded in relatively tight formation for 75 m, with the scout periodically darting in and out of the lead. Just beyond 75 m, however, the scout alone advanced 3 m to the NW. The ant swarm remained behind and started exploring in expanding circles. The scout remained by herself, making wide, looping movements for 3.5 minutes By the end of this period, many of the raiders near the back of the swarm had turned around and were slowly heading homeward. Finally, the scout rejoined the swarm, and both scout and workers returned to the mixed nest.

Although the scout clearly functions to arouse nestmates and direct them to target colonies, additional observations show that she is not the only individual capable of group recruitment. On several occasions in the oak-woodland habitat (in which *Formica* nests are concealed beneath rocks and leaf litter), the *Polyergus* scout leading a raid stopped advancing. As was typical in such situations, the raiding workers scattered and began exploring the area. Although there were several sites of intense *Polyergus* activity, no *Formica* emerged. After 5 minutes the scout and raiders gradually started back to their home nest. We then attempted to determine whether the lack of raid success was due to faulty scouting (as the early literature had suggested) or to the ability of *Formica* to block their nest entrance. When we excavated several small rocks in the area where the *Polyergus* had explored, we indeed exposed several *F. gnava* adults and pupae. The few remaining *Polyergus* that contacted these *Formica* became highly aroused, ran toward their returning nestmates and recruited about 150 *Polyergus* raiders back to the *Formica* nest. By this time the original scout was 4 m from the *Formica* colony and was therefore also part of the back-recruited swarm.

The Stimulus for Panic-Alarm by *Formica*

A slave raid by *Polyergus* immediately elicits panic-alarm in the target colony, as adult workers (and the queen) of *Formica* flee in all directions from their nest. This alarm behavior could be caused by a pheromone secreted by *P. breviceps* or by chemical and tactile interactions among the raided *Formica* workers. To distinguish between these possibilities, 3 small plastic boxes (each 18 cm × 12.5 cm × 6 cm high) were placed in a row on a board of plywood, interconnected by plastic tubes (3 cm long × 2 cm diameter). The first box had a 2-cm diameter entrance hole near the base of the front wall, and the third box contained an identical exit hole near the base of its rear wall. One hundred adult *Formica* and 100 pupae were placed in each of the 3 boxes, and each group promptly sequestered their brood in one corner of their respective containers. This 3-box unit was placed 3 m in front of a *Polyergus* raid swarm (Topoff, Cover & Jacobs, 1989).

On each of 3 trials, hundreds of *Polyergus* workers were recruited into the first box immediately after the raiders located the entrance hole and encountered the resident *Formica*. This interaction set off a wave of panic-alarm inside the first box, as workers of Formica picked up pupae and scurried into the second box. The arrival of aroused, brood-carrying *Formica* was immediately accompanied by a similar panic-alarm of the ants in box #2, and by the end of the first minute (since the initial entrance of *Polyergus* into box #1), the arousal had spread into the third box, where the exit hole permitted adult *Formica* to remove their pupae from the test chambers. Although the *Polyergus* were in hot pursuit, it is important to note that, for each of the three trials, more than 50 *Formica* pupae had been removed from the last box before the *Polyergus* raiders had entered even the second box.

To determine whether olfactory stimuli alone could have in-duced this panic-alarm in *Formica*, 100 adult *Formica* workers and 100 pupae were placed in another small plastic box. The middle section of one wall of the box was replaced with wire mesh, and the bottom of the box was coated with Fluon. This small box was then placed on top of 2 wooden struts, inside a second, larger plastic box (30 cm × 16 cm × 9 cm high). The base of the front wall of this box contained a large opening (12 cm long × 2 cm high). Finally, 100 *Formica* pupae (but no adult workers) were also placed on the floor of the large box.

To begin a test, the 2-box set was placed 3 m in front of a *Polyergus* raid swarm, with the opening facing the advancing workers. The combination of wooden struts and Fluon prevented physical con-tact between raiding *Polyergus* and the resident *Formica* adults in the smaller box. Chemical secretions by *Polyergus*, however, could reach the *Formica* adults by diffusion through the wire-mesh wall.

On each of 3 trials, hundreds of *Polyergus* individuals swarmed into the large box, became highly aroused by the presence of *Formica* pupae and started active recruitment. All the *Formica* pupae were removed in less than 3 min., and all were immediately transported back to the *Polyergus* nest. But despite the immense activity level of the *Polyergus* raiders, the adult *Formica* workers in the smaller box were not aroused. They remained tightly clustered, and no brood carrying was observed.

This test series was repeated with 100 adult *Formica* and 100 pupae placed in both boxes. In each of these 3 trials, the adult *Formica* in the large outer box exhibited panic-alarm (with brood carrying) as soon as contact was made with entering *Polyergus*. More importantly, however, the *Formica* adults in the inner container also became highly aroused, as they too picked up their pupae and ran excitedly around the small box. These results suggest that secretions from adult *Formica* adults elicit panic-alarm in nestmates.

Colony Takeover by *Polyergus* Queens

The life of a socially parasitic ant colony is fraught with uncertainties, not the least of which is how a parasitic queen can invade and take over a host colony. In laboratory studies with the myrmicine ant *Harpagoxenus canadensis*, invading queens employ two contrasting mechanisms for invading host nests (Stuart, 1984). In the "active" method, the *Harpagoxenus* queen moves quickly and erratically inside the nest, eliciting severe biting from resident workers. Because this high level of aggression also causes host workers to pick up their brood and abandon the nest, actively invading queens usually obtain only a fraction of the colony's brood. During "passive" invasion, by contrast, the parasitic queen advances slowly, antennating all residents encountered. Because this furtive technique generates little alarm behavior, passive queens are able to commandeer the entire host brood and even be adopted by a large portion of the resident adult workers.

For *Polyergus*, several field studies have clarified how parasitic queens locate host colonies in the first place. In the eastern species *P. lucidus*, for example, mated queens frequently return to a *Polyergus* nest, and follow slave raids to target nests of *Formica* (Cool-Kwait & Topoff, 1984; Marlin, 1968; Talbot, 1968). In our Arizona field site, we found that queens of *P. breviceps* have an even more efficient mechanism for reaching *Formica* colonies. Instead of a mating flight, most winged *Polyergus* queens run in the slave raid and copulate within the advancing swarm. The queens stop running momentarily, attract males with a pheromone produced in the mandibular glands, mate and shed their wings and then promptly continue with the raiding workers to the target nest (Topoff & Greenberg, 1988). By mating in slave-raid swarms, fertile *Polyergus* queens arrive at raided colonies of *Formica* whose workers and queen are scattered across the substrate. Such disorganization in the target colony could facilitate the queen's penetration into *Formica* colonies.

If the techniques used by *Polyergus* queens to locate *Formica* colonies are unique, the adaptations necessary for successful adoption by the foreign species are nothing less than spectacular. A recent study on the parasitic myrmicine ant *Leptothorax kutteri* disclosed a pheromone used by queens to facilitate colony usurpation of its host *L. acervorum* (Allies, Bourke & Franks, 1986). Originating in the queen's Dufour's gland, this exocrine secretion caused host workers to attack each other, possibly by superseding nestmate recognition within the colony. Similarly, our laboratory studies (Topoff et al., 1988) showed that *Polyergus* queens also use a pheromone from their enlarged Dufour's glands to reduce aggression during attempted colony takeover.

During host-queen killing, the *Polyergus* queens spent a median of 26 minutes (range = 11–34 minutes, N = 13) biting the resident *Formica* queens. Observations with a dissecting microscope disclosed that between bouts of biting, the attacking queen's mandibles are widely separated, as she continually touches her extruded hypopharynx to the dead queen. Before the death of the *Formica* queen, all interactions between *Formica* workers and the *Polyergus* queen are hostile. But within minutes after the resident queen's death, worker attacks on the *Polyergus* queen are replaced with bouts of grooming. By the end of each successful adoption, the *Polyergus* queen straddles the *Formica* brood and stands with her legs fully extended, thus exposing all surfaces to the grooming *Formica* workers.

Perhaps the most intriguing result of our colony-adoption tests was that the success rate for colony takeover by *P. breviceps* queens dropped from 79% to 12% when no *Formica* queen was present. In the queenless condition, aggression by *Formica* workers toward the *Polyergus* queen typically resulted in her death. This suggests that during the act of biting the *Formica* queen, the parasitic *Polyergus* queen acquires one or more *Formica*-queen chemicals. And this, in turn, led to the prediction that *Polyergus* queens will attack and bite *Formica* queens, even if they are dead and motionless.

To test this hypothesis, we collected newly mated queens of *P. breviceps*, and colonies of *Formica gnava* and *F. occulta*. Immediately prior to testing, we froze a *F. gnava* queen for 5 minutes, thawed her for an additional 5 minutes, and placed her in a nest containing 15 workers and 15 pupae. A single *Polyergus* queen was then introduced into the arena surrounding the nest chamber. For each of 5 replications, we recorded the number of attacks (discrete bites) the *Polyergus* queen made on the *F. gnava* queen, the accumulated attacking time and time until colony takeover was completed. Our criterion for successful takeover was met when: (1) the *Polyergus* queen was being groomed by colony workers; and (2) the queen was sitting on or near the brood pile.

Each of the 5 *Polyergus* queens successfully took over *F. gnava* colonies containing dead queens, with a median time to takeover of 140 minutes. The median accumulated attack time by the *Polyergus* queen was 23 minutes which was not significantly different from the attack time (median = 26 minutes) recorded for our previous study using living queens of *Formica* ($X2 = 0.8$, $df = 1$, $P > 0.30$). In each instance the *Polyergus* queen attacked the dead *F. gnava* queen repeatedly, directing bites over the *Formica* queen's head, thorax, legs, petiole and gaster. After biting and piercing the dead queen, each invading *Polyergus* queen used her hypopharynx to lick all bitten surfaces, followed by long bouts of grooming. The entrance of the *P. breviceps* queen caused some of the *F. gnava* workers to leave the nest and scatter in the adjoining arena. Those that stayed in the nest aggressively attacked the intruding *Polyergus* queen, often with prolonged bites to her limbs. After the *Polyergus* queen pierced the *Formica* queen, worker behavior changed abruptly from attack to antennation and grooming (Topoff & Zimmerli, 1993).

Colony Takeover by Queens and the Evolution of Social Parasitism

We recently proposed an evolutionary explanation for social parasitism in *Polyergus*, by focusing on colony takeover by queens and then deriving slave and territorial raids from the establishment of interspecific bonds through olfactory imprinting (Topoff, 1990; Topoff & Zimmerli, 1993). Our scenario begins with a newly mated, preparasitic queen overrunning a colony of *Formica*, killing the resident queen and being adopted by the workers. An objection to this idea was raised by Buschinger (1986), who noted that invading a foreign colony is a risky venture. In response, we offer 3 arguments. First, colony founding even by free-living queens is hazardous, because the vast majority of queens are routinely killed by workers from neighboring colonies and by numerous arthropod, reptilian and avian predators. Second, the success rate of queens would be increased if invasions were restricted to incipient nests containing only a few, small-sized workers. Queen killing and usurpation of incipient nests have recently been documented for *P. breviceps* (Topoff & Mendez, 1990). Third, and most important, any queen successfully adopted by even a tiny nest of a related species would achieve a great reproductive advantage, because the attendant host workers could immediately offer some degree of protection, while simultaneously caring for her brood. Obviously, this strategy would only be successful if queen killing transferred the necessary chemicals from the host to the invading queen, and if the food (and other ecological) requirements of the parasite and host were similar (which also explains why socially parasitic ants are typically closely related to their hosts).

Our evolutionary hypothesis led to the question of whether a *Polyergus* queen could successfully take over a colony of a different *Formica* species if given the opportunity to acquire the odor of the heterospecific queen? As a test, we placed each of 7 newly mated *Polyergus* queens from colonies containing *F. gnava* slaves in a nest of the closely related species *F. occulta*, containing 15 adult workers, 15 pupae and one living *F. occulta* queen. Conversely, we placed 7 *Polyergus* queens from colonies containing *F. occulta* slaves in *F. gnava* nests containing 15 adult workers, 15 pupae and one *F. gnava* queen. For each trial, we recorded the number of attacks by the *Polyergus* queen on the resident *Formica* queen, the accumulated attacking time, and time until the takeover criterion was met. We observed behavioral interactions between the *Polyergus* queen and *Formica* workers continuously for 2 hours, then approximately every hour until the takeover criterion was met (or until the *Polyergus* queens died).

Not surprisingly, there were 10 unsuccessful takeovers, and all were marked by noticeably more acute aggression (involving repeated biting, rolling and gaster flexing) than seen in sympatric takeovers. Four of the *Polyergus* queens showed no interest in attacking the *Formica* queens: they encountered the *Formica* queens, antennated them and then either ignored or avoided them. The deaths of the unsuccessful queens were largely the result of constant attacks by the *Formica* workers prior to the killing of the *Formica* queen. Nevertheless, 4 of the queens were successful in meeting the takeover criterion (range = 50 minutes–3 days). As in our previous tests, in which *Polyergus* attacked only *Formica* of the same species found in their home nest, the successful takeovers were marked by no aggression immediately after the *Polyergus* queen killed the *Formica* queen (Topoff & Zimmerli, 1993). In one case, the *Polyergus* queen died several days after takeover, most likely from injuries sustained during pretakeover aggression.

It is clear that during the evolution of obligatory social parasitism, a queen must have entered a nest of a closely related species, killed the resident queen and usurped the workers and brood, much as we demonstrated in this last test series. After successful adoption, our preparasitic *Polyergus* queen lays eggs in the host nest, and her brood is cared for by the resident *Formica* workers. As the *Polyergus* brood matures, olfactory imprinting occurs between the two species, so that adult workers of *Polyergus* and *Formica* eventually treat each other as conspecifics (not to mention nestmates). Evidence for imprinting during the ontogeny of both free-living and parasitic ants is now well documented (Isingrini, Lenoir & Jaisson, 1985; Le Moli & Mori, 1985). Now the stage is set for the origin of slave raids. If an adult *Polyergus* forager from this nest encounters a foreign colony of the same species, she promptly recruits nestmates to a territorial raid. During these aggressive conflicts, both colonies experience high mortality (Topoff et al., 1984). More important, however, is: what happens when this *Polyergus* forager locates a nest of *Formica*? Because *Polyergus* now identifies individuals of *Formica* as belonging to its own species, another territorial raid is initiated. Thus from the standpoint of *Polyergus*, according to this theory, slave raids and territorial raids are functionally equivalent. But because free-living workers of *Formica* are easily aroused and exceedingly susceptible to the effects of formic acid (and other alarm pheromones), they react to these raids with panic-alarm instead of overt aggression (Topoff et al., 1989). They bolt from the nest, leaving their pupal brood vulnerable to unchallenged retrieval. As successive raided broods of *Formica* are reared in the mixed species nest and join the workforce, our ancestral *Polyergus* could easily slide in the direction of facultative parasitism. And when the *Polyergus* workers and queen eventually lose the ability to care for themselves or their own brood, the relationship with *Formica* becomes obligatory. In conclusion, the theory presented here puts the evolutionary first step on colony usurpation by queens. Its advantage over previous theories is that it is unnecessary to postulate an independent mechanism for raiding with brood capture. A successful colony adoption by species capable of olfactory imprinting will automatically lead to a behavioral equivalence between intraspecific territorial and interspecific slave raids.

Brood Parasitism and the Question of Comparison

The meteoric rise of animal behavior studies in the United States, starting in the decade 1955–1965, was due primarily to the influence of the European ethologists. Following the traditions established by Lorenz, Tinbergen and von Frisch, the study of species-typical behavior was firmly rooted in biology; indeed, comparative anatomy provided both the theoretical and methodological tools necessary for ethologists to elucidate behavioral homologies and eventually the phylogeny of behavior. Schneirla (1952) appreciated the contribution of this approach but argued that comparative psychology should not be so narrowly based. The American ethologists imported a whole arsenal of ethological terminology—instinct, sign stimuli and releasers, to name but a few. Because these evolutionary-oriented biologists focussed primarily on (phylogenetic) changes in the adaptive patterns of behavior in adult organisms, concepts such as "instinct" were routinely applied to species as diverse as amebas, ants and aardvarks. The problem, of course, was that many people, especially nonscientists, concluded that similar descriptive terminology implied comparably similar mechanisms underlying species-typical behavior in widely divergent taxa. Schneirla certainly agreed with ethologists that natural selection had "produced" remarkably similar adaptations to solve common problems. For example, when individuals live in relatively stable societies, we typically find that social bonds have been established, that patterns of communication exist for behavioral integration and that behavioral specialization (in the form of division of labor) is common. Indeed, we are not surprised to find these analogous adaptive outcomes in slime molds, army ants or troops of baboons.

Brood parasitism, in which eggs are laid in the nest of another species and the young reared by the hosts, is clearly a form of interspecific social behavior that has evolved in insects, fishes and birds (Baba, Nagata & Yamagishi, 1990). Being a successful parasite also requires the solution of a set of common problems, including: (1) which species to parasitize, (2) how to locate the host's nest, (3) how to avoid being driven out of the nest and (4) how to ensure that the eggs layed will be reared and not destroyed. The table illustrates these analogous adaptations for brood parasitism between *Polyergus* and cuckoos (as reported by Davies & Brooke, 1988).

Table 1. Comparison of Principal Behavioral Adaptations Between Cuckoos and *Polyergus*

Cuckoo	Polyergus
Parasitizes four host species	Parasitizes three host species (in some habitats)
Individual parasitic female specializes on one host species	Individual colony specializes on one host species
Parasite locates host nest by observing host build nest	Parasitic female locates host nest by running in slave raid or by searching alone on substrate
Parasitic female enters nest when host parents are absent	Parasitic female enters host when host queen is gone (during slave raid) or when host queen is present
Parasitic female removes one egg and lays her own	Parasitic queen kills host queen, then lays eggs
Parasitic chick fed same food as host chicks	Parasite brood and workers fed same food as host workers

Cuckoo	Polyergus
Host selection occurs via imprinting (i.e., female selects host of same species in whose nest she hatched)	Host selection occurs via olfactory imprinting (i.e., parasitic queen selects same species of *Formica* present in her nest during eclosion)

Schneirla's studies of learning in ants and rodents did not imply a belief that ants are ancestral to rats. Nor was it based on the simplistic notion that the use of tactile and olfactory cues by both ants and rats represented a fundamentally equivalent instinctive mechanism for these stimuli. Schneirla was equally interested in differences in the pattern of maze learning, attributed, of course, to corresponding differences in neural organization between the insect and rodent. In commenting on his own comparative studies of learning, Bitterman reports a wide range of similarities in the performance of honeybees and vertebrates. But he also correctly reminds us that many of the similarities in outcomes may indeed reflect the operation of convergent but quite different functional principles. This, of course, is exactly the position taken when we compare brood parasites in ants and birds. The similarities are undoubtedly convergent, but behavioral convergences are as much a product of evolution as are behavioral homologies. The differences in the mechanisms and processes underlying these matching adaptations are also the products of evolution, as they reflect qualitatively different neuroanatomical specializations, ecological requirements and divergent historical origins. Indeed, it is by concentrating equally on these two aspects of animal behavior that comparative psychology stands apart from, and in some respects is more comprehensive than, classical ethology.

References

Allies, A., Bourke, A., & Franks, N. (1986). Propaganda substances in the cuckoo ant *Leptothorax kutteri* and the slave-maker *Harpagoxenus sublaevis. Journal of Chemical Ecology, 12,* 1285–1293.

Alloway, T. M. (1980). The origins of slavery in Leptothoracine ants. *American Naturalist, 115,* 247–261.

Baba, R., Nagata, Y., & Yamagishi, S. (1990). Brood parasitism and egg robbing among three freshwater fish. *Animal Behaviour, 40,* 776–778.

Buschinger, A. (1986). Evolution of social parasitism in ants. *Trends in Ecology and Evolution, 1,* 155–160.

Cool-Kwait, E., & Topoff, H. (1984). Raid organization and behavioral development in the slave-making ant *Polyergus lucidus Mayr. Insectes Sociaux, 31,* 361–374.

Davies, N. B., & Brooke, M. (1988). Cuckoos versus reed warblers: adaptations and counteradaptations. *Animal Behaviour, 36,* 262–284.

Emery, C. (1908). Osservazioni ed esperimenti sulla Formica Amazzone. *Rendicanto delle Sessioni della Reale Accademia delle Scienze, dell' Istituto di Bologna, 12,*49–62.

Harkness, R. D., & Maroudas, N. G. (1985). Central place foraging by an ant (*Cataglyphis bicolor Fab.*): a model of searching. *Animal Behaviour, 33,* 916–928.

Harman, J. R. (1968). Some aspects of the ecology of the slave-making ant, *Polyergus lucidus. Entomological News, 79,* 217–223.

Huber, P. (1810). *Recherchez sur les Meours des Fourmis Indigenes.* Paschoud, Paris.

Isingrini, M., Lenoir, A., & Jaisson, P. (1985). Preimaginal learning as a basis of colony-brood recognition in the ant *Cataglyphis cursor. Proceedings of the National Academy of Sciences, 82,* 8545–8547.

Le Moli, F., & Mori, A. (1985). The influence of the early experience of worker ants on enslavement. *Animal Behaviour, 33,* 1384–1387.

Marlin, J. C. (1968). Notes on a new method of colony formation employed by *Polyergus lucidus Mayr. Transactions of the Illinois State Academy of Sciences, 61,* 207– 209.

Schneirla, T. C. (1952). A consideration of some conceptional trends in comparative psychology. *Psychological Bulletin, 49,* 559–597.

Stuart, R. J. (1984). Experiments on colony foundation in the slave-making ant Harpagoxenus canadensis M.R. Smith (Hymenoptera; *Formicidae). Canadian Journal of Zoology, 62,* 1995–2001.

Talbot, M. (1967). Slave raids of the ant *Polyergus lucidus. Psyche, 74,* 299–313.

Talbot, M. (1968). Flights of the ant *Polyergus lucidus Mayr. Psyche, 75,* 46–52.

Topoff, H. (1990). The evolution of slave-making behavior in the parasitic ant genus *Polyergus. Ethology Ecology and Evolution, 2,* 284–287.

Topoff, H., Bodoni, D., Sherman, P., & Goodloe, L. (1987). The role of scouting in slave raids by *Polyergus breviceps* (Hymenoptera: *Formicidae). Psyche, 94,* 261–270.

Topoff, H., Cover, S., Greenberg, L., Goodloe, L., & Sherman, P. (1988). Colony founding by queens of the obligatory slave-making ant *Polyergus breviceps:* the role of the Dufour's gland. *Ethology, 78,* 209–218

Topoff, H., Cover, S., & Jacobs, A. (1989). Behavioral adaptations for raiding in the slave-making ant, *Polyergus breviceps. Journal of Insect Behavior, 2,* 545–556.

Topoff, H., & Greenberg, L. (1988). Mating behavior of the socially parasitic ant *Polyergus breviceps:* the role of the mandibular glands. *Psyche, 95,* 81–87.

Topoff, H., LaMon, B., Goodloe, L., & Goldstein, M. (1984). Social and orientation behavior of *Polyergus breviceps* during slave-making raids. *Behavioral Ecology and Sociobiology, 15,* 273–279.

Topoff, H., & Mendez, R. (1990). Slave raid by a diminutive colony of the socially parasitic ant, *Polyergus breviceps. Journal of Insect Behavior, 3,* 819–821.

Topoff, H., & Zimmerli, E. (1993). Colony takeover by a socially parasitic ant, *Polyergus breviceps:* the role of chemicals obtained during host-queen killing. *Animal Behaviour, 46,* 479–486.

Wehner, R., & Srinivasan, M. V. (1981). Searching behavior of desert ants, genus Cataglyphis (*Formicidae, Hymenoptera). Journal of Comparative Physiology, 142,* 315–338.

Wheeler, W. M. (1916). Notes on some slave-raids of the western amazon ant (*Polyergus breviceps Emery). Journal of the New York Entomological Society, 24,* 107–118.

Wilson, E. O. (1971). *The insect societies.* Cambridge, Mass.: Harvard University Press.

Wilson, E. O. (1975). *Leptothorax duloticus* and the beginnings of slavery in ants. *Evolution, 29,* 108–119.

The Role of Workers and Queens in the Colony-Member Recognition Systems of Ants

Are There Any Differences that Predispose Some Kinds of Ants to Social Parasitism?

Thomas M. Alloway

Ants live in complex societies called "colonies." In most species, each colony contains 3 basic kinds of adult individuals: queens, workers and males. Queens are reproductive females. Workers are more or less sterile females that construct and defend the nest, forage for food and rear the offspring of the colony's queen(s). Males are parthenogenetically produced reproductives whose only biological role is to mate with queens; they perform little or no labor on behalf of the colony. In most species, queens are morphologically different from workers. Queens are larger, initially have wings that they shed after their mating flight and possess well-developed ovaries and a receptaculum to store sperm. Workers are the wingless individuals that one commonly encounters outside the nest. Their ovaries are either absent or less developed than those of queens, and they lack a receptaculum in most species.

Ants go through the 4-phase life cycle that is typical of insects that undergo complete metamorphosis. The egg hatches to produce a larva that eats, grows and sheds its skin several times before becoming a pupa. In about half the ant species, the larva spins a cocoon in which the pupa develops into an adult. In the other half, the larva pupates without spinning a cocoon. Ant larvae are helpless creatures that are incapable of moving around under their own power or getting their own food. They require constant care. The workers must not only feed them, they must keep them clean and move them about from place to place in the nest to insure that they are exposed to regimes of temperature and humidity that will facilitate their healthful and timely development. The eggs, larvae and pupae in an ant colony are collectively referred to as the colony's "brood."

Because ant workers are sterile (or nearly so), the most important social relationship in a worker's life is that between the worker and the queen(s) whose offspring the worker rears. If the worker rears the offspring of her mother or some other close relative with whom she shares many genes, her investment of labor results in the production of many replicates of the worker's own genes even if the worker produces no offspring of her own (Hamilton, 1964). However, if the worker rears the offspring of an unrelated queen, her labor is wasted in the replication of genes that she does not share. For this reason we might expect that robust mechanisms would have evolved to ensure that workers invest their labor in the offspring of kin only (Hamilton 1964).

A mechanism that has this effect under many circumstances is related to the fact that, in most species, ant colonies are closed societies. By a closed society, I mean that strangers are excluded. The workers of a particular colony labor cooperatively with one another, but members of other similar colonies are treated as enemies. If workers from another colony of the same or a closely related species are encountered outside the nest, they may be attacked; and the colony is ready to invest great effort and, if need be, to expend the lives of many workers to prevent outsiders from entering the nest. Because all the colony's workers are the offspring of the colony's queen(s), the maintenance of a closed society generally has the effect of insuring that workers are related to the larvae that they rear.

Yet the phenomena of social parasitism show that the system is not foolproof. Social parasites are species that induce workers of other so-called host species to rear the offspring of parasite-species queens. As members of another species, the parasite-species offspring are most definitely not the close relatives of host-species workers. There are three major kinds of social parasitism in which the labor of host-species workers is exploited:

1.   Temporary Parasitism. Temporary-parasite queens found new colonies parasitically. New colonies of most nonparasitic ant species are founded by one or more newly inseminated young queens that find a suitable but unoccupied nest site and rear a first batch of workers. A second nonparasitic mode of colony foundation is called "budding": One or more inseminated young queens depart from

their parental colony in the company of some workers and start a new colony with the workers' assistance. Temporary-parasite queens enter a host-species colony, replacing the host-species queen(s) either by killing them or inducing the host-species workers to kill their own queen(s). The parasites then begin laying eggs that the host-species workers rear to produce parasite-species workers. This form of parasitism is called "temporary" because the host-species workers are not replaced when they die. Thus, as the colony grows, the number of parasite-species workers increases and the number of host-species workers declines, with the eventual result being a pure parasite-species colony (Hölldobler & Wilson, 1990).

2.   Inquilinism. Most species involved in this form of parasitism have no worker caste. All the adult members of the parasite species are either queens or males. After mating, a parasite queen enters a host-species colony and somehow secures adoption. The queens of most inquiline parasite species allow the host-species queen(s) to survive. The parasitized colony thus continues to produce host-species workers, but all or most of the reproductives that it produces belong to the parasite species (Hölldobler & Wilson, 1990).

3.   Slavery (= "dulosis"). In most species, slave-maker queens found colonies in essentially the same way as temporary parasites, by entering a host-species colony and replacing the host-species queen(s). The principal difference between the two forms of parasitism arises from the fact that the slave-maker workers are morphologically and behaviorally specialized for raiding. Instead of (or in species of *Formica sanguinea* species group in addition to) performing ordinary worker-ant functions, the slave-maker workers attack other host-species colonies, kill or drive off the adult inhabitants of raided nests and carry captured brood back to the slave-maker nest. There young host-species workers that mature from the captured brood (so-called slaves) form a social attachment to the slave makers and perform their repertoire of worker behaviors in the service of the slave-maker colony (Hölldobler & Wilson, 1990).

These phenomena raise questions about the means by which ant colonies maintain their closure and about whether some means may be more effective than others in resisting the depredations of social parasites.

The maintenance of closed societies is typical of most, but not all, kinds of ants. The main exceptions are the so-called unicolonial species. Most of these species maintain dense, local populations consisting of numerous, somewhat interconnected nests that as a group may contain millions of workers and thousands of queens (e.g., certain species of the genus *Formica* [Cherix, 1980; Higashi, 1983; Mabelis, 1986]). Another demographic correlate of failure to maintain a closed society may be the presence of large numbers of queens in colonies (e.g., Keller & Passera, 1989; Morel, Vander Meer & Lofgren, 1990). An interesting exception is *Leptothorax retractus*, a species that lives in colonies that usually contain less than 100 workers and only one or a very few queens. Stuart (in press) has recently reported that *L. retractus* is quite unaggressive toward workers from other conspecific nests. *L. retractus'* lack of aggressiveness stands in stark contrast to the tightly closed societies maintained by the closely related sympatric *L.* species B and many other *Leptothorax* species (e.g., Alloway et al., 1982; Provost, 1979, 1985). Interestingly, although *L.* species B is the host to two workerless inquiline parasites (Heinze, 1989; Heinze & Alloway, 1991) and both *L.* species B and a third sympatric species *L.* species A are frequent hosts to the slavemaker *Harpagoxenus canadensis*, *L. retractus* is rarely enslaved and hosts no known inquilines. Recently, Leighl (1992) observed under laboratory conditions that *L. retractus* slaves in *H. canadensis* nests fail to form a normal social attachment to the slave makers. When an unenslaved *L. retractus* colony is placed in an arena with an *H. canadensis* colony containing *L. retractus* slaves, many *L. retractus* workers abandon the slave makers and join the unenslaved *L. retractus* colony.

Among the majority of social insect species that do live in closed colonies, discrimination between colony members and others is based on chemical cues (Hölldobler & Wilson, 1990). Although the nature of the chemical cues has not been identified with certainty, the most likely candidates are complex hydrocarbons found on the surface of the exoskeleton (Bonavita-Cougourdan et al., 1987; Clément et al., 1987; Morel & Vander Meer, 1988; Nelson et al., 1980; Vander Meer & Wojcik, 1982). These chemical cues are often referred to as the "colony odor."

Three important questions regarding colony odors are:

1. Which colony members produce the chemical cues?
2. How do workers acquire the capacity to distinguish them?
3. Which members of the colony are distinguished?

These questions have been addressed with the greatest thoroughness in two very different kinds of ants: carpenter ants of the genus *Camponotus* and several species of the genus *Leptothorax*. The two genera are phylogenetically about as distantly related as any two kinds of ants can be. *Camponotus* belongs to the subfamily Formicinae: *Leptothorax*, to the subfamily Myrmicinae. These two subfamilies are located at diametrically opposite sides of the ant phylogenetic tree (Hölldobler & Wilson, 1990). Carpenter ants are large insects that live in populous colonies that frequently contain several thousand workers. *Leptothorax* species are minute ants that live in small colonies which generally contain less than 100 workers. In addition, whereas species of *Camponotus* are virtually never subject to social parasitism, the temperate-zone species of *Leptothorax* in which nestmate recognition has been most thoroughly studied are hosts to sizable retinues of social parasites.

Three broad sources of colony-member recognition cues have been postulated. The cues may be produced by workers, derived from queens or imparted to the insects by constituents of the environment (e.g., food or nesting material) (Carlin & Hölldobler, 1986). As we shall see, all three components play a role in both *Camponotus* and *Leptothorax*.

To determine the role of these constituents in colonies, investigators often create mixed colonies by rearing workers of different species or workers derived from different colonies of the same species together from pupae. The fact that workers derived from different mutually hostile colonies will live together peacefully if reared together during the first days of their adult lives shows that learning plays a crucial role in colony-member recognition. This basic fact about the psychology of colony-member recognition in ants has been known since Huber's (1810) pioneering study of ant slavery.

Carlin and Hölldobler (1986) studied the relative importance of queen-, worker- and food-derived cues in the nestmate-recognition system of several species of *Camponotus*, whose mature colonies ordinarily contain only one queen (monogyny) and occupy only one nest site (monodomy). The study involved the creation of 4 series of small experimental colonies: (1) colonies containing a young queen and workers of 2 species, (2) colonies containing a young queen and workers derived from 2 colonies of the same species, (3) queenless worker groups composed of workers derived from two colonies of the same or different species and (4) pairs of worker groups each of which contained workers derived from two colonies and between which a single queen was repeatedly switched. Some experimental colonies composed of workers derived from the same stock colonies were fed different diets to determine the potential contribution of food odors to colony-member identification. After the workers had matured, arena tests were run to determine whether pairs of workers from different experimental colonies would accept each other or fight. The results of these arena tests permitted inferences about the relative contributions of queen-, worker-, and environment-derived recognition cues. The principal finding was that queen-derived cues were more important than worker-derived cues in colonies containing a queen, but that worker-derived cues still enabled queenless worker groups to detect aliens. The effect of food-derived cues was small. These investigators also examined the time course of acceptability for young *Camponotus* workers in conspecific colonies. Workers were highly acceptable up to about 5 days of age, after which their acceptability declined.

Carlin and Hölldobler (1986), in the study just described, employed small experimental colonies containing fewer than 10 workers. Carlin and Hölldobler (1987) examined larger experimental colonies containing approximately 190 workers. In these larger colonies, the relative importance of queen- and worker-derived cues depended upon the physiological status of the queens. Queen-derived cues still predominated over worker-derived cues in colonies where the queens were fully fertile. However, worker-derived cues were more important in colonies containing a less fertile queen. Nevertheless, Carlin and Hölldobler (1988) showed that even nonfertile virgin queens are capable of imparting some queen-derived cues to workers.

Hamilton's (1964) theory of kin selection as a prime mover for the evolution of highly complex social behavior (so-called eusocial[1] behavior) in ants, wasps and bees predicts that workers should be able to perceive their degree of genetic relatedness to other members of their colony and that the greatest amount of cooperation should occur among the closest relatives. Carlin, Hölldobler and Gladstein (1987) examined this possibility in experimental colonies of *Camponotus floridanus* in which the workers were derived from two different stock colonies. Half the experimental colonies contained queens; the other half did not. No differences in the frequency of food exchange or grooming were observed between the groups of related and unrelated workers. However, unrelated workers antennated one another more often than related workers. In queenless nests, where aggression among the workers sometimes occurred, more aggressive acts were directed at nonkin than at kin. These results were interpreted as offering some rather equivocal support for Hamilton's (1964) theory.

Stuart has elucidated the means by which workers discriminate between fellow colony members and strangers in *Leptothorax ambiguus*, *L. curvispinosus*, and *L. longispinosus*, three closely related species that nest primarily in old, partly hollow acorns and hickory nuts in eastern North America. Colonies of all three of these species are closed, sometimes contain more than one functional queen (facultative polygyny) and sometimes occupy more than one nest (facultative polydomy) (Alloway et al., 1982). In one series of studies (Stuart, 1985, 1987a, 1992), *Leptothorax* worker pupae were removed from their parental nests and allowed to eclose in isolation, after which the young workers were maintained in isolation for varying periods of time ranging from a few hours to 157 days before being introduced either to their parental nest, another conspecific nest or an allospecific nest. Very young *Leptothorax* workers less than 4 hours after emergence from the pupa were often accepted into both conspecific and allospecific colonies. Between 4 and 60 hours, workers were often accepted into alien conspecific colonies but were no longer acceptable in allospecific nests (Stuart, 1985, 1992). However, workers of all ages were acceptable in their parental colony. The fact that isolated workers remain acceptable indefinitely in their parental colony shows that individual *Leptothorax* workers produce colony-specific recognition cues; and the simplest explanation of the gradually decreasing acceptability of young workers in allospecific and conspecific colonies other than their parental colony is that the colony-specific recognition cues that ultimately cause workers to be rejected develop gradually during the first few days of a young worker's life.

Leptothorax queens may also produce colony-specific recognition cues. By dissecting all the queens in a sample of polygynous nests of *L. ambiguus*, *L. curvispinosus* and *L. longispinosus*, Alloway et al. (1982) discovered that colonies of these species sometimes adopt inseminated young queens. Stuart, Gresham-Bisset and Alloway (1993) studied this queen-adoption process by observing newly mated young queens after they had been introduced to either their parental colony or another conspecific colony. The result was that 16 of 27 queens introduced to their parental colonies (59%), but only one of 41 introduced to other conspecific colonies (2%), were accepted. Although it is unclear why a substantial number of colonies rejected their own newly mated young queens, the much higher acceptance rates for young queens introduced to parental, as opposed to other, colonies suggests that queens produce colony-specific identifiers.

In other experiments, Stuart (1985, 1988a) showed that, when *Leptothorax* workers live together in groups, they somehow share one another's recognition cues. One kind of experiment involved placing a worker pupa in an allospecific colony and allowing the worker to emerge from the pupa and be adopted into the alien colony. When these workers were subsequently removed from their adoptive colony and returned to their parental colony, they were attacked. A second kind of experiment involved switching varying numbers of worker pupae between pairs of colonies. These studies showed that if a sufficiently large number of worker pupae were reciprocally exchanged between two colonies, the colonies lost the capacity to distinguish themselves from each other. Thus, individual workers not only produce colony-specific recognition cues; groups of *Leptothorax* workers that are living together share their individual cues and base colony-member recognition on these shared identifiers.

Since the capacity of *Leptothorax* colonies to recognize and reject strangers is not affected by the number of queens that the colony contains (Stuart, 1991), the role of worker-derived recognition cues is thought to be more important than that of queen-derived cues in these *Leptothorax* species. However, Provost (1979, 1985) has shown that queen-derived cues play a more important role in certain other *Leptothorax* species.

Environmental cues may also play an important role in *Leptothorax* nestmate recognition. Stuart (1987b) compared the reactions of workers in *L. curvispinosus* test nests to intruders from other *L. curvispinosus* nests that had been collected either very close to the test nest (0.09–1.87 m), farther from the test nest but at the same collection site (1.52–4.65 m) or at another collection site (7 km). Two sets of tests were made, the first 2–4 weeks after the nests had been collected and set up in identical artificial nests and the second 13–17 weeks after collection. Ants in the test nests were sometimes unaggressive toward workers from other nests collected very close to the test nest at the same collection site, a result that further documents the finding of Alloway et al. (1982) that some *L. curvispinosus* colonies are polydomous. However, the principal finding of interest was that, although the high degree of aggressiveness that ants in the test nests manifested toward workers from nests collected at another site did not change over time, aggressiveness displayed toward workers from distal nests collected at the same site significantly decreased over time. Stuart (1987b) interpreted this reduction in aggressiveness as being indicative of the role of environmentally derived colony-member identification cues that dissipate over time when colonies are cultured in identical artificial nests and fed identical diets. Future laboratory experiments should test this inference in studies designed to show whether colony cohesion can be disrupted when colony halves are cultured in different kinds of nests and/or fed different diets.

Recently, my laboratory has been examining the role of queens in the colony-member recognition system of *Leptothorax ambiguus*. Alloway and Ryckman (1991) assessed whether *L. ambiguus* workers are always able to recognize queens to which they are genetically related (a finding that would support Hamilton's [1964] theory of kin selection) or whether social attachment to queens is learned. In a preliminary experiment, we simply demonstrated that *L. ambiguus* colonies do in fact recognize their own queens. On the day before the experiment, all queens were removed from 20 *L. ambiguus* nests. The next day, one of their own queens and a queen from another colony were introduced, one at a time, to each nest. As Figure 1 shows, *L. ambiguus* colonies bit and stung their own ("concolonial") queens less and groomed them more than they directed these behaviors toward queens from other ("allocolonial") colonies.

Alloway and Ryckman's (1991) second experiment assessed whether queen recognition is intrinsic or learned. A total of 64 *L. ambiguus* colonies were randomly divided into 4 groups of 16 colo-

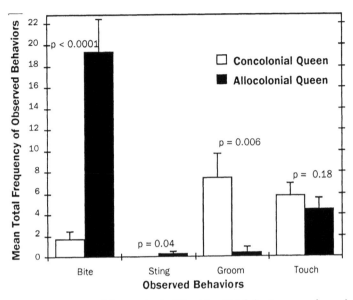

Figure 1. Mean total frequencies (+ SE) with which behaviors were observed when field-collected L. ambiguus *colonies were exposed to a concolonial and allocolonial* L. ambiguus *queen.*

nies each; and 5 worker pupae from each colony were isolated and allowed to mature to adulthood in an artificial nest with a queen. In one group, the queens with which the pupae were isolated came from the same colony as the pupae. In the other 3 groups, the pupae were isolated with either a queen from another *L. ambiguus* colony, an *L. longispinosus* queen or a *Protomognathus americanus* slave-maker queen. Fourteen days after the last worker had eclosed in each isolation group, the queen with which the young workers had been living was removed from the group; and the following day the group was observed for 20 minutes after the reintroduction of their familiar queen and for another 20 minutes after the introduction of an unfamiliar queen from the worker group's parental colony. As Figure 2 shows, worker groups that had matured with a concolonial queen made no distinction between familiar and unfamiliar queens from that same colony. How-

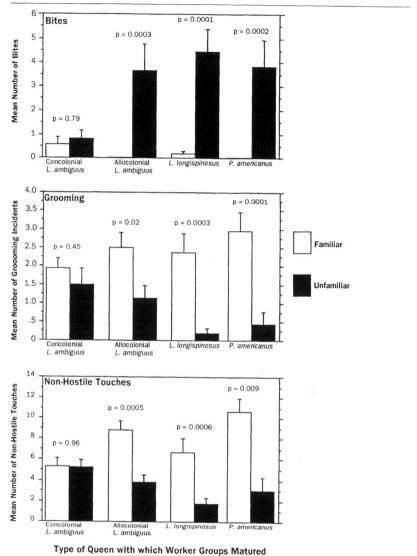

Figure 2. Mean total number of bites, grooming episodes and nonhostile touches (+SE) observed when groups of young L. ambiguus workers were exposed to the "familiar" queen with which they had lived since adult maturation or to an unfamiliar L. ambiguus queen from their parental colony.

ever, groups that had matured with either an allocolonial *L. ambiguus* queen, an *L. longispinosus* queen or a *Protomognathus* slave-maker queen bit the familiar queen less and groomed and nonhostilely touched her more than they directed these behaviors at unfamiliar queens from their parental colony. All these results are compatible with the hypothesis that *L. ambiguus* workers learn to become socially attached to any queen with which they spend the first days of their adult lives.

Alloway (under editorial review) investigated whether *L. ambiguus* queens contribute identifying cues to workers with which they are living. Thirty *L. ambiguus* nests were selected for designation as "experimental nests" on the basis of containing 2 or more queens, 30 or more workers and 15 or more worker pupae. Each of these experimental nests was paired with a "queen-source nest" that had been selected on the criteria of coming from a different collection site, containing approximately the same number of workers as the experimental colony with which it was paired and containing at least one queen. At the beginning of the experiment, 2 groups of 5 worker pupae were removed from each experimental colony and placed in isolation nests, one group with a queen from the experimental colony in question and the other group with a queen from that experimental colony's matched queen-source colony. Fourteen days after the maturation of the last worker pupa in each isolation group, 2 workers were randomly selected for service as test workers and removed from the group; and the next day one of these workers was introduced to its parental colony while the other was placed in the queen-source colony with which its parental colony had been matched. The results of the introductions are shown in Figure 3.

1.   Workers in queen-source colonies bit test workers reared with queen-source colony queens less than they bit test workers reared with experimental-colony queens.

2.   Workers in experimental colonies groomed test workers reared with experimental-colony queens more than they groomed test workers reared with queens from queen-source colonies.

3.   Workers in experimental colonies nonhostilely touched test workers that had been living with experimental-colony queens more than they touched test workers reared with queen-source colony queens, and workers in queen-source colonies nonhostilely touched test workers that had been living with queen-source colony queens more than they touched test workers reared with experimental-colony queens.

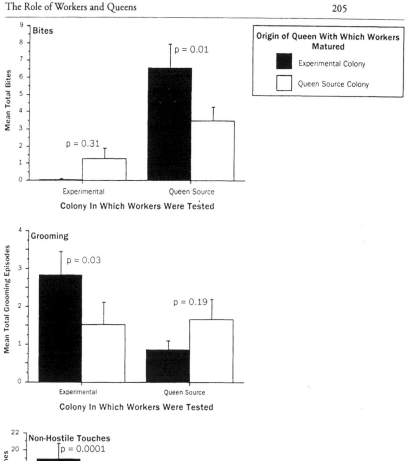

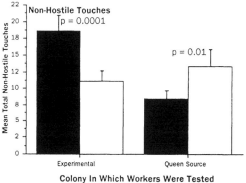

Figure 3. Mean total number of bites, grooming episodes and nonhostile touches (+SE) directed by experimental and queen-source colonies toward test workers that had matured and been living with queens from either experimental or queen-source colonies.

All these results are compatible with the hypothesis that queens impart identifying cues to workers with which they live.

In a second experiment, Alloway (under editorial review) determined whether workers impart identifying cues to queens. Thirty *L. ambiguus* nests were selected for designation as "experimental nests" on the basis of containing 2 or more dealate queens and 10 or more worker pupae. Each of these experimental nests was paired with a "pupa-source" nest that had been selected on the criteria of coming from a different collection site and containing more than 5 worker pupae. At the beginning of the experiment, 2 queens from each experimental colony were isolated in separate artificial nests. Five worker pupae from the experimental colony were placed in the nest with one queen, while 5 worker pupae from that experimental colony's pupa-source colony were placed in the nest with the other queen. Fourteen days after the last worker had matured in each isolation group, the queen was removed; and the next day the queen was reintroduced to the experimental colony from which she had originated. The experimental hypothesis that workers impart colony-identification cues to queens with which they are living predicts that queens that have been living with workers derived from their own colony should be more readily accepted there than queens that have been living with workers derived from another colony. The results of these queen introductions are depicted in Figure 4. Queens that had been living with young workers derived from their own experimental colonies were bitten less, groomed more and nonhostilely touched more than queens that had been living with workers from pupa-source colonies. These findings are compatible with the hypothesis that workers do in fact impart colony-identification cues to queens.

In summary, recent work from my laboratory indicates that *L. ambiguus* workers learn social attachments to queens much as they learn social attachments to one another and that *L. ambiguus* workers and queens reciprocally share one another's colony-identification cues. These findings extend Stuart's (1985, 1987a, 1987b, 1988a) earlier findings with regard to worker-derived cues and strongly suggest that *Leptothorax* workers and queens share the same colony-member recognition system.

At this point, it is possible to speculate about what form of

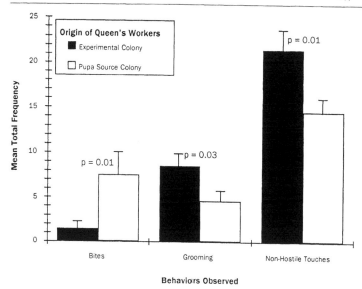

Figure 4. Mean total number of bites, grooming episodes and nonhostile touches (+SE) directed by experimental colonies toward experimental-colony queens that had been living with young workers derived either from experimental or pupa-source colonies.

learning is involved in *Leptothorax* colony-member identification. In several papers, Stuart (1987a, 1987b, 1988a) and I (Alloway & Ryckman, 1991; Alloway & Keough, 1990) have stated that *Leptothorax* workers learn to identify fellow colony members on the basis of colony-characteristic mixtures of chemical cues. While true, such statements might suggest that *Leptothorax* workers possess a sophisticated capacity to learn to discriminate the combination of chemical cues characteristic of their own colony from combinations characterizing other colonies. However, we need not assume that *Leptothorax* workers learn gestalt-like combinations of cues; and there is good reason to believe that they probably do not. For one thing, since colonies of these species frequently occupy more than one nest and workers and queens living in different nests probably contact one another only occasionally, there is reason to doubt whether all members of a colony would possess an identical combination of recognition cues upon which a gestalt-learning system could be based. Moreover, one of Stuart's (1985, 1987a, 1992) most important findings is that several-month-old workers that have been isolated from the pupa stage are still highly and uniquely acceptable in their parental colony. These socially isolated workers, which have had no opportunity to acquire colony-identification cues from nestmates and which thus presumably bear on their bodies only their own intrinsic identifiers, would presumably be rejected if colony-member recognition were based on the possession of some colony-specific cue combination that included cues from all, or even a significant proportion of, the colony's inhabitants. These facts make it highly unlikely that *Leptothorax* workers learn to identify nestmates on the basis of gestalt-like cue combinations. More probably, colony-member identification in these species is based on a much simpler learning process resembling habituation (Alloway, 1972; Mazur, 1990), and Stuart (1987c, 1988b) has alluded to this possibility. According to this view, as young workers mature, they habituate to the recognition cues found on the workers and queens that they encounter and thereby learn not to attack ants bearing those cues. At the same time, if a maturing ant's recognition cue is unfamiliar to its nestmates, older individuals habituate to the new worker's cue. Thereafter, individuals are accepted as members of the same colony if they carry no unfamiliar recognition cues; and individuals are rejected and attacked with increasing intensity as the number and quantity of any unfamiliar cues that they bear increases.

Another category of colony members whose acceptance or rejection is of interest is the brood. Brood acceptance is particularly important as it relates to the phenomenon of ant slavery. Slave makers kidnap brood from raided host-species colonies and carry it home to their slaves. If a slave-maker colony is to be successful, the slave workers must not only accept and rear the parasite-species brood, they must also accept and rear captured broods. Brood acceptance has been extensively studied in carpenter ants of the genus *Camponotus*, in the acorn-dwelling *Leptothorax* species of eastern North America that are hosts to three slave-making parasites (*Protomognathus americanus*, *Leptothorax duloticus*, and *L. pillagens*) and in several species of the genus *Formica*, some species of which are enslaved by sanguinary (*F. sanguinea* species group) and amazon (the genus *Polyergus*) slave makers.

Jaisson (1975) showed that *Formica polyctena* workers learn to tend brood of any species to which they are exposed during the first days of their adult lives and that subsequently they treat brood of any species to which they were not exposed as prey. *Formica* workers not exposed to brood of any species during early adult life subsequently consume brood of all species. Since this learning takes place during a sensitive period that lasts for about a week following adult emergence, Jaisson (1975) and Jaisson and Fresneau (1978) called the learning process involved "imprinting." LeMoli and his associates (LeMoli & Passetti, 1977, 1978; LeMoli & Mori, 1982) subsequently extended Jaisson's results to *F. lugubris* and *F. rufa*; and Goodloe and Topoff (1987) found that the *Formica* slaves of *Polyergus* slave makers accept more brood of their own than of another host species. However, LeMoli (1978) showed that social experience after the sensitive period can modify the behavior of *F. rufa* workers to some extent. Workers that had been deprived of brood during their sensitive period initially destroyed brood from their own colony. However, when placed with a group of experienced workers and pupae from their colony, the initially brood-deprived workers learned to tend pupae instead of destroying them. A learning mechanism at least superficially similar to the imprinting process that accounts for the acquisition of brood-acceptance behavior in *Formica* may also be operative in amazon slave makers of the genus *Polyergus*. *Polyergus lucidus* colonies never contain but one species of slaves; and the slave makers consistently raid colonies of the same species as the slaves that they already possess (Goodloe, Sanwald & Topoff, 1987). *P. lucidus* queens also regularly found new colonies by invading *Formica* nests of the same species as the slaves in their parental colony (Goodloe & Sanwald, 1985).

Alloway (1982) examined the effect of enslavement on the pupa-acceptance behavior of *Leptothorax* workers by comparing the pupa-acceptance behavior of enslaved workers kept in the presence of *Protomognathus americanus* (= *Harpagoxenus americanus*) slave-maker workers, enslaved workers from which the slave makers had been removed and unenslaved workers. In every condition, the *Leptothorax* workers accepted and maintained a greater proportion of conspecific than allospecific host brood. However, something about the experience of enslavement and the environment of slave-maker colonies enhanced brood-acceptance behavior. Enslaved workers kept and reared a greater proportion of allospecific host-species brood than unenslaved workers, and the effect was especially pronounced among the groups kept with slave-maker workers.

Subsequently, Hare and Alloway (1987) examined the process by which young *Leptothorax* workers learn their preference for conspecific brood. Groups of *L. ambiguus* and *L. longispinosus* workers were allowed to emerge from pupae in the presence of brood of their own species, brood of the other species or without brood. Subsequently, workers that had eclosed with conspecific brood manifested a nonexclusive preference for brood of their own species, while workers that had eclosed with allospecific brood or with no brood tended brood of both species nonpreferentially. Carlin et al. (1987) independently obtained the same result with *Camponotus floridanus* and *C. tortuganus* carpenter ants. These results for *Leptothorax* and *Camponotus* are fundamentally different from those obtained with *Formica* (Jaisson, 1975; Jaisson & Fresneau, 1978; LeMoli, 1978; LeMoli & Mori, 1982; LeMoli & Passetti, 1977, 1978):

1.   Whereas young *Formica* workers acquire a preference for tending brood of any species to which they are exposed during the first days of their adult lives, young *Camponotus* and *Leptothorax* workers are capable of learning a preference only for brood of their own species.

2.   Whereas young *Formica* workers subsequently eat brood of any species to which they were not exposed during the first days of their adult lives, *Camponotus* and *Leptothorax* workers manifest preferences much less dramatically. *Camponotus* and *Leptothorax* workers that have been conditioned to conspecific brood and are manifesting a preference for tending brood of their own species merely retrieve conspecific brood on average more quickly than they retrieve allospecific brood; and they rear greater proportions of conspecific than allospecific brood. Nevertheless, nonpreferred brood is retrieved and substantial proportions of nonpreferred brood are reared.

These differences suggest that *Camponotus* and *Leptothorax* workers learn their preferences in a different way than *Formica* workers. Instead of manifesting a learning process similar to the imprinting mechanism that various authors (Jaisson, 1975; Jaisson & Fresneau, 1978; LeMoli, 1978; LeMoli & Mori, 1982; LeMoli & Passetti,

1977, 1978) have attributed to *Formica*, *Camponotus* and *Leptothorax* workers probably acquire their preferences for conspecific brood through a simpler learning process resembling sensitization (Alloway, 1972; Mazur 1990).

One of the interesting peculiarities of social parasitism in ants is that socially parasitic species are by no means randomly distributed with respect to either geography or taxonomy. Eighty-nine percent of the 218 species of social parasites listed by Hölldobler and Wilson (1990) occur in the North Temperate Zone. Among the 11 ant subfamilies recognized by Hölldobler and Wilson (1990), 82% of social parasites occur in just 2: the Formicinae and the Myrmicinae (Hölldobler & Wilson 1990). Moreover, the parasites are not randomly distributed even within these subfamilies. Only 17 of the 127 genera of Myrmicinae and 8 of the 49 genera of Formicinae contain host species. Since temporary, inquiline and slave-making parasites are generally more closely related to their hosts than to any other nonparasitic forms (Emery, 1909; Wilson, 1971; Hölldobler & Wilson, 1990), these facts mean that there is very likely something special about certain kinds of ants that favors the evolution of social parasitism within a small number of taxonomic groups. Since the vast majority of ant species live in the tropics but the vast majority of social parasites and hosts live in the North Temperate Zone, there may also be something special about that part of the world.

As an example of combined taxonomic and geographic restriction, consider ant slavery. Slavery has probably evolved independently 9 or 10 times. Among the Formicinae, it probably evolved at least twice in the genus *Formica* (to produce the *Formica sanguinea* group of sanguinary slave makers and the genus *Polyergus* of amazon slave makers) and once in the genus *Proformica* (to produce the slave-making genus *Rossomyrmex*). Among the Myrmicinae, it probably evolved at least 5 or 6 times in the genus *Leptothorax* (once or twice to produce slave-making species of *Leptothorax* and another 4 times to produce the parasite genera *Chalepoxenus*, *Epimyrma* [= *Myrmoxenus*], *Harpagoxenus*, and *Protomognathus*) and once in the genus *Tetramorium* (to produce the parasite genus *Strongylognathus*). All known slave makers and their hosts occur in the North Temperate Zone.

The title of this chapter asks whether certain kinds of colony-member recognition systems may be more susceptible than others to exploitation by social parasites. In other words, are colony-member recognition systems the "something special" that facilitate the evolution of social parasitism in certain taxonomic groups. Unfortunately, colony-member recognition has been studied in so few taxonomic groups that no definitive answer can be given to this question at the present time.

Apart from the imprinting-like learning process by means of which *Formica* workers learn to recognize brood (Jaisson, 1975; Jaisson & Fresneau, 1978; LeMoli, 1978; LeMoli & Mori, 1982; LeMoli & Passetti, 1977, 1978), little is known about colony-member recognition in the genus *Formica*, which has given rise to two major groups of slave makers, a host of temporary parasites and at least one workerless inquiline (Hölldobler & Wilson, 1990).

Among the leptothoracines, we have extensive information about colony-member recognition in three species (*Leptothorax ambiguus*, *L. curvispinosus* and *L. longispinosus*) that are hosts to three slave-makers (*L. duloticus*, *L. pillagens* and *Protomognathus americanus*) and one workerless inquiline (*L. minutissimus*, a parasite of *L. curvispinosus*) (Hölldobler & Wilson, 1990; Alloway, unpublished observations; S. Cover, personal communication). As we have seen, in these ants, workers and probably also queens produce colony-specific recognition cues, workers and queens that are living together share one another's cues and adult colony members are recognized as any individuals that bear no unfamiliar cues.

As we have also seen, there is a psychological difference between the ways *Leptothorax* and *Formica* workers acquire preferences for tending conspecific brood. However, since the genus *Formica* is as subject to parasitism as *Leptothorax* and since *Camponotus*, a genus with virtually no social parasites, shares the *Leptothorax* mode of acquiring preferences for conspecific brood, the *Leptothorax* means of acquiring brood species preferences is probably not the key to understanding why social parasitism has evolved so often in the genus *Leptothorax*.

Among *Camponotus*, adult colony-member recognition seems to be basically similar to that found in *Leptothorax* in that chemical cues derived from queens, workers and the environment all play a role. The main difference is that chemical cues emanating from

Camponotus queens are relatively more important than is the case in *Leptothorax*. Conceivably, this seemingly small difference is crucial. If so, that fact will become apparent when colony-member recognition is studied more fully in other groups of ants—both those that are and those that are not subject to social parasitism. However, I see no reason why having the queen as the principal source of the "colony odor" should protect a species from social parasitism. It is easy to imagine a slave maker in which the queen was the principal source of colony-identifying chemical cues for both her own workers and the colony's slaves. Similarly, one can imagine a scenario in which inquiline parasite queens could gain easy access to host colonies by being "odorless" and allowing themselves to be marked by colony-identifying substances derived from the host queen(s).

Another approach to the question of the role of colony-member recognition in social parasitism involves trying to understand how the process works in mixed colonies containing workers of more than one species. So little is known about colony-member recognition in pure *Formica* colonies that there is little point in speculating about how the process might work in sanguinary or amazon slave-maker colonies. However, we know enough about the *Leptothorax* hosts of the slave-makers *Protomognathus americanus*, *L. duloticus* and *L. pillagens* to make some educated guesses. As we have seen, *Leptothorax* workers belonging to the hosts of these slave makers produce their own colony-identifying cues; and workers living together share one another's cues and learn to recognize as nestmates any individuals that lack unfamiliar cues. Since the slave makers obtain new slaves as a result of capturing host-species brood, it is easy to understand why young host-species workers acquire their social attachment to the slave-maker colonies; and the readiness with which even unenslaved *Leptothorax* workers accept and tend brood from other colonies and other host species (Hare & Alloway, 1987) makes it easy to understand why colonies of these three slave makers often contain slave workers of more than one species.

Nevertheless, recent investigations of *Protomognathus americanus* have revealed that this slave maker has evolved some interesting modifications to and ways of disrupting the colony-member recognition system of its *Leptothorax* hosts. In what is probably a very ancient relationship between this slave maker and its hosts, the hosts have also evolved special means of dealing with *P. americanus*. Alloway and Hare (1989) studied the reaction of host-species workers to *P. americanus* brood. *L. longispinosus* workers were allowed to mature from pupae in the presence of either *P. americanus* larvae, *L. longispinosus* larvae or no larvae at all. When subsequently tested, *L. longispinosus* workers in all three conditions manifested a nonexclusive preference for retrieving and tending the slave-maker brood. This result contrasts with the findings of Hare and Alloway (1987) in parallel experiments, described above, that showed that *Leptothorax* workers reared in the presence of conspecific brood subsequently preferred brood of their own species over brood of another host species. In other words, Alloway and Hare's (1989) results with *P. americanus* brood show that something about the slave-maker larvae is sufficiently attractive to overcome the *Leptothorax* host-species workers' normal preference for tending conspecific brood.

During raids, *P. americanus* disrupts the colony-member recognition system of target colonies that it is about to attack. Slave raids are initiated by "scouts," slave-maker workers that find a host-species target nest and then return home to lead a raiding party. Host-species workers that discover a slave-maker scout near their nest usually attack the intruder. *L. duloticus*, *L. pillagens* and *H. canadensis* scouts counterattack and frequently kill hostile host-species workers encountered near the target nest (Alloway, 1979, unpublished observations; Stuart & Alloway, 1982, 1983). *P. americanus* scouts, in contrast, allow target-nest defenders to drag them around. However, any target-nest worker that has touched a *P. americanus* scout is immediately attacked by its nestmates. When several target-nest workers discover and attack a *P. americanus* scout, widespread fighting may break out among the target-colony workers. When the *P. americanus* scout eventually escapes and returns with a raiding party, the target colony is already disorganized and less able to resist the raiders (Alloway, 1979; Del Rio Pesado & Alloway, 1984). Similar tactics are employed by certain other leptothoracine social parasites, including the slave-maker *Harpagoxenus sublaevis* and the workerless inquiline *Leptothorax kutteri* (Allies, Bourke & Franks, 1986; Buschinger, 1968, 1974). In both these latter species, the disruptive substance is produced in the Dufour's gland (Allies, Bourke & Franks, 1986).

P. americanus colonies also "brand" their slaves. Alloway and Keough (1990) reared *L. longispinosus* ant workers from pupae as slaves in *P. americanus* nests and subsequently tested the acceptability of these enslaved workers by introducing them either to the *L. longispinosus* colony from which they had been taken as pupae or to another unenslaved *L. longispinosus* colony. In addition, as controls, the acceptability of unenslaved *L. longispinosus* workers was tested by introducing them into their own and other *L. longispinosus* colonies. Half the slaves lived in nests containing slave-maker adults until just prior to testing, while the other half lived in slave-maker nests from which the adult slave makers had been removed one week before testing. Slaves that had been living with slavemakers just prior to testing were attacked more than unenslaved workers and more than slaves which had been living in nests from which the slave makers had been removed, but slaves that had been living without slave makers were attacked no more than unenslaved workers. These findings suggest that *P. americanus* chemically marks its slaves. The function of this marking is unknown, but it might be related to the fact that *P. americanus* raids are relatively nonviolent affairs. Instead of killing large numbers of the adults in raided nests, as *L. duloticus, L. pillagens* and *Harpagoxenus canadensis* raiders do, *P. americanus* raiders incite panic that causes *Leptothorax* colonies to abandon their nests without suffering many casualties (Alloway, 1979, unpublished observations; Alloway & Del Rio Pesado, 1984; Del Rio Pesado & Alloway, 1984; Stuart & Alloway, 1982, 1983). As a consequence, the colonies from which *P. americanus* derives its slaves are likely to survive and persist in close proximity to *P. americanus* nests. Perhaps slave branding is a way of preventing slaves from returning to their nearby parental colonies. The possibility that unbranded slaves might be able to return to their parental colonies is not as farfetched as it sounds. Stuart (1985, 1988a) found that, although *Leptothorax* workers that had been adopted into other *Leptothorax* host-species colonies were at first attacked after being returned to their parental nests, they were rarely killed and frequently eventually adopted.

The hosts of *P. americanus* have also adapted to the slave maker. Alloway (1990) compared the reactions of *L. longispinosus* colonies after the introduction of a *P. americanus* and *L. ambiguus* workers. More *L. longispinosus* workers bit and stung *P. americanus* than *L.*

ambiguus intruders; and after the introduction of a single slave maker, the *L. longispinosus* colonies frequently carried brood outside their nest and sometimes evacuated the nest entirely, responses that were virtually never seen after the introduction of an *L. ambiguus* worker. *L. longispinosus* colonies thus take special defensive measures against intrusions by slave makers that they seldom or never take against their territorial competitor *L. ambiguus*. So far, this is the only known instance of "enemy recognition" on the part of a host to a social parasite, although species whose colonies are frequently raided by army ants are known to take special defensive measures against the incursions of these predators (Droual, 1984; LaMon & Topoff, 1981).

These last-cited studies show that, although the *Leptothorax* colony-member recognition system provides a good substrate for the evolution of social parasitism, long-lasting host-parasite interactions may lead to small modifications of the system in ways that serve the interests of both the parasites and their hosts.

Note

1. Eusocial insects live in societies where there is reproductive division of labor (i.e., some individuals reproduce and others are nonreproductive workers), complex care of the young and an overlap of generations (i.e., the workers are the offspring of the reproductives).

References

Allies, A. B., Bourke, A. F. G., & Franks, N. R. (1986). Propaganda substances in the cuckoo ant *Leptothorax kutteri* and the slave-maker *Harpagoxenus sublaevis*. *Journal of Chemical Ecology, 12*, 1285–1293.

Alloway, T. M. (1972). Learning and memory in insects. *Annual Review of Entomology, 17*, 43–56.

Alloway, T. M. (1979). Raiding behavior of two species of slave-making ants, *Harpagoxenus americanus* (Emery) and *Leptothorax duloticus* Wesson (Hymenoptera: Formicidae). *Animal Behavior, 27*, 202–210.

Alloway, T. M. (1982). How the slave-making ant *Harpagoxenus americanus* affects the pupa-acceptance behavior of its slaves. In M. D. Breed, C. D. Michener, & H. E. Evans (eds.), *The biology of social insects*, (pp. 261–265). Boulder, Colo.: Westview Press.

Alloway, T. M. (1990). Slave-species ant colonies recognize slavemakers as enemies. *Animal Behaviour, 39*, 1218–1220.

Alloway, T. M. (In press). The role of queens in the colony-member recognition system of the ant *Leptothorax ambiguus* Emery. *Animal Behaviour*.

Alloway, T. M., Buschinger, A., Talbot, M., Stuart, R., & Thomas, C. (1982). Polygyny and polydomy in three North American species of the ant genus *Leptothorax* Mayr (Hymenoptera: Formicidae). *Psyche, 89*, 249–274.

Alloway, T. M., & Del Rio Pesado, M. G. (1984). Behavior of the slave-making ant, *Harpagoxenus americanus* (Emery), and its host species under seminatural laboratory conditions (Hymenoptera: Formicidae). *Psyche, 90,* 425–436.

Alloway, T. M., & Hare, J. F. (1989). Experience-independent attraction to slave-maker ant larvae in host-species ant workers (*Leptothorax longispinosus*; Hymenoptera: Formicidae). *Behaviour, 110,* 93–105.

Alloway, T. M., & Keough, G. (1990). Slave marking by the slave-making ant *Harpagoxenus americanus* (Emery) (Hymenoptera: Formicidae). *Psyche, 97,* 55–64.

Alloway, T. M., & Ryckman, D. (1991). Learned social attachment to queens in *Leptothorax ambiguus* Emery ant workers. *Behaviour, 118,* 235–243.

Bonavita-Cougourdan, A., Clément, J.-L., & Lange, C. (1987). Nestmate recognition: the role of cuticular hydrocarbons in the ant *Camponotus vagus* Scop. *Journal of Entomological Science, 22,* 1–10.

Buschinger, A. (1968). Untersuchungen an *Harpagoxenus sublaevis* Nyl. (Hymenoptera: Formicidae), III: Kopula, koloniegründung, raubzüge. *Insectes Sociaux, 15,* 89–104.

Buschinger, A. (1974). Experimente und beobachtungen zur gründung und entwicklung neuer sozietäten der sklavenhaltenden ameise *Harpagoxenus sublaevis* (Nyl.). *Insectes Sociaux, 21,* 381–406.

Carlin, N. F., & Hölldobler, B. (1983). Nestmate and kin recognition in interspecific mixed colonies of ants. *Science, 222,* 1027–1029.

Carlin, N. F., & Hölldobler, B. (1986). The kin recognition system of carpenter ants (*Camponotus* spp.) I. Hierarchical cues in small colonies. *Behavioral Ecology and Sociobiology, 19,* 123–134.

Carlin, N. F., & Hölldobler, B. (1987). The kin recognition system of carpenter ants (*Camponotus* spp.) II. Larger colonies. *Behavioral Ecology and Sociobiology, 20,* 209–217.

Carlin, N. F., & Hölldobler, B. (1988). Influence of virgin queens on kin recognition in the carpenter ant *Camponotus floridanus* (Hymenoptera: Formicidae). *Insectes Sociaux, 35,* 191–197.

Carlin, N. F., Halpern, R., Hölldobler, B., & Schwartz, P. (1987). Early learning and the recognition of conspecific cocoons by carpenter ants (*Camponotus* spp.). *Ethology, 75,* 306–316.

Carlin, N. F., Hölldobler, B., & Gladstein, D. S. (1987). The kin recognition system of carpenter ants (*Camponotus* spp.) III. Within-colony discrimination. *Behavioral Ecology and Sociobiology, 20,* 219–227.

Cherix, D. (1980). Note préliminaire sur la structure, la phénologie et le régime alimentaire d'une super-colonie de *Formica lugubris* Zett. *Insectes Sociaux, 27,* 226–236.

Clément, J.-L., Bonavita-Cougourdan, A., & Lange, C. (1987). Nestmate recognition and cuticular hydrocarbons in *Camponotus vagus* Scop. In J. Eder & H. Rembold (eds.), *Chemistry and biology of social insects* (pp. 473–474). Munich: Verlag J. Peperny.

Del Rio Pesado, M. G., & Alloway, T. M. (1984). Polydomy in the slave-making ant, *Harpagoxenus americanus* (Emery) (Hymenoptera: Formicidae). *Psyche, 90,* 151–162.

Droual, R. (1984). Anti-predator behaviour in the ant *Pheidole desertorum*: The importance of multiple nests. *Animal Behaviour, 32,* 1054–1058.

Emery, C. (1909). Über den ursprung der dulotischen, parasitischen und myrmekophilen ameisen. *Biologisches Centralblatt, 29,* 352–362.

Goodloe, L., & Sanwald, R. (1985). Host specificity in colony-founding by *Polyergus lucidus* queens (Hymenoptera: Formicidae). *Psyche, 92,* 297–302.

Goodloe, L., Sanwald, R., & Topoff, H. (1987). Host specificity in raiding behavior of the slave-making ant *Polyergus lucidus*. *Psyche, 94,* 39–44.

Goodloe L. P., & Topoff, H. (1987). Pupa acceptance by slaves of the social-parasitic ant *Polyergus*. *Psyche, 94,* 293–302.

Hamilton, W. D. (1964). The genetical evolution of social behavior, I and II. *Journal of Theoretical Biology, 7,* 1–52.

Hare, J. F., & Alloway, T. M. (1987). Early learning and brood discrimination in leptothoracine ants (Hymenoptera: Formicidae). *Animal Behaviour, 35,* 1720–1724.

Heinze, J. (1989). *Leptothorax wilsoni* n. sp., a new parasitic ant from eastern North America (Hymenoptera: Formicidae). *Psyche, 96,* 49–61.

Heinze, J., & Alloway, T. M. (1991). *Leptothorax paraxenus* n. sp., a workerless social parasite from North America (Hymenoptera: Formicidae). *Psyche, 98,* 195–206.

Heinze, J., Stuart, R. J., Alloway, T. M., & Buschinger, A. (1992). Host specificity in the slave-making ant, *Harpagoxenus canadensis* M. R. Smith. *Canadian Journal of Zoology, 70,* 167–170.

Higashi, S. (1983). Polygyny and nuptial flight of *Formica (Formica) yessensis* Forel at Ishikari Coast, Hokkaido, Japan. *Insectes Sociaux, 30,* 287–297.

Hölldobler, B., & Wilson, E. O. (1990). *The ants.* Cambridge, Mass.: Belknap/Harvard.

Huber, P. (1810). *Recherches sur les Mœurs des Fourmis Indigènes.* Geneva: J. J. Paschoud.

Jaisson, P. (1975). L'imprégnation dans l'ontogénèse des comportements de soins aux cocons chez la jeune fourmi rousse (*Formica polyctena* Forst). *Behaviour, 52,* 1–37.

Jaisson, P., & Fresneau, D. (1978). The sensitivity and responsiveness of ants to their cocoons in relation to age and methods of measurement. *Animal Behaviour, 26,* 1064–1071.

Keller, L., & Passera, L. (1989). Influence of the number of queens on nestmate recognition and attractiveness of queens to workers in the Argentine ant, *Iridomyrmex humilis* (Mayr). *Animal Behaviour, 37,* 733–740.

LaMon, B., & Topoff, H. (1981). Avoiding predation by army ants: Defensive behaviours of three ant species of the genus *Camponotus. Animal Behaviour, 29,* 1070–1081.

Leighl, A. (1992). Factors affecting host specificity and habitat selection in the slave-making ant *Harpagoxenus canadensis* M. R. Smith and its host species. Unpublished M.Sc. thesis, University of Toronto.

LeMoli, F. (1978). Social influence on the acquisition of behavioral patterns in the ant *Formica rufa* L. *Bolletino di Zoologia, 45,* 399–404.

LeMoli, F., & Mori, A. (1982). The effect of early learning on recognition, acceptance and care of cocoons in the ant *Formica rufa* L. *Bolletino di Zoologia, 49,* 93–97.

LeMoli, F., & Passetti, M. (1977). The effect of early learning on recognition, acceptance and care of cocoons in the ant *Formica rufa* L. *Atti della Società Italiana di Scienze Naturali e del Museo Civile di Storia Naturale, Milano, 118,* 49–64.

LeMoli, F., & Passetti, M. (1978). Olfactory learning phenomena and cocoon nursing behaviour in the ant *Formica rufa* L. *Bolletino di Zoologia, 45, 389–397.*

Mabelis, A. A. (1986). Why do young queens fly? (Hymenoptera, Formicidae). *Proceedings of the Third European Congress of Entomology* (pp. 24–29) Amsterdam.

Mazur, J. E. (1990). *Learning and behavior.* 2nd ed. Englewood Cliffs, N.J.: Prentice-Hall.

Morel, L., & Vander Meer, R. K. (1988). Ontogeny of nestmate recognition in the red carpenter ant *(Camponotus floridanus):* behavioral and chemical evidence for the role of age and social experience. *Behavioral Ecology and Sociobiology, 22,* 175–183.

Morel, L., Vander Meer, R. K., & Lofgren, C. S. (1990). Comparison of nestmate recognition between monogyne and polygyne populations of *Solenopsis invicta* (Hymenoptera: Formicidae). *Annals of the Entomological Society of America, 83,* 643–647.

Nelson, D. R., Fatland, C. L., Howard, R. W., McDaniel, C. A., & Blomquist, G. J. (1980). Re-analysis of the cuticular methylalkanes of *Solenopsis invicta* and *S. richteri. Insect Biochemistry, 10,* 409–418.

Provost, E. (1979). Étude de la fermeture de la société de fourmis chez diverses espèces de *Leptothorax* et chez *Camponotus lateralis* (Hyménoptères:

Formicidae). *Comptes Rendus de l'Académie des Sciences, Paris*, ser. D, *288*, 429–432.

Provost, E. (1985). Étude de la fermeture de la société chez les fourmis, I: Analyse des interactions entre ouvrières de sociétés différentes, lors de rencontres expérimentales, chez des fourmis du genre *Leptothorax* et chez *Camponotus lateralis* Ol. *Insectes Sociaux, 32*, 445–462.

Stuart, R. J. (1985). *Nestmate recognition in leptothoracine ants: Exploring the dynamics of a complex phenomenon*. Unpublished Ph.D. thesis, University of Toronto.

Stuart, R. J. (1987a). Individual workers produce colony-specific nestmate recognition cues in the ant, *Leptothorax curvispinosus*. *Animal Behaviour, 35*, 1062–1069.

Stuart, R. J. (1987b). Transient nestmate recognition cues contribute to a multicolonial population structure in the ant, *Leptothorax curvispinosus*. *Behavioral Ecology and Sociobiology, 21*, 229–235.

Stuart, R. J. (1987c). Nestmate recognition in leptothoracine ants: Testing Fielde's progressive odor hypothesis. *Ethology, 76*, 116–123.

Stuart, R. J. (1988a). Collective cues as a basis for nestmate recognition in polygynous leptothoracine ants. *Proceedings of the National Academy of Science* (USA), *85*, 4572–4575.

Stuart, R. J. (1988b). Development and evolution in the nestmate recognition systems of social insects. In G. Greenberg & E. Tobach (eds.), *Evolution of Social Behavior and Integrative Levels* (pp. 175–195). Hillsdale, N.J.: Lawrence Erlbaum Associates.

Stuart, R. J. (1991). Nestmate recognition in leptothoracine ants: testing the effects of queen number, colony size, and species of intruder. *Animal Behaviour, 42*, 277–284.

Stuart, R. J. (1992). Nestmate recognition and the ontogeny of acceptability in the ant, *Leptothorax curvispinosus*. *Behavioral Ecology and Sociobiology, 30*, 403–408.

Stuart, R. J. (In press). Differences in aggression among sympatric, facultatively polygynous *Leptothorax* ant species. *Animal Behaviour.*

Stuart, R. J., & Alloway, T. M. (1982). Territoriality and the origin of slavery in leptothoracine ants. *Science, 215*, 1262–1263.

Stuart, R. J., & Alloway, T. M. (1983). The slave-making ant, *Harpagoxenus canadensis* M. R. Smith, and its host species, *Leptothorax muscorum* (Nylander): Slave raiding and territoriality. *Behaviour, 85*, 58–90.

Stuart, R. J., Gresham-Bisset, L., & Alloway, T. M. (1993). Queen adoption in the polygynous and polydomous ant, *Leptothorax curvispinosus*. *Behavioral Ecology, 4*, 244–262.

Vander Meer, R. K., & Wojcik, D. P. (1982). Chemical mimicry in the myrmecophilous beetle *Myrmecaphodius excavaticollis*. *Science, 218*, 806–808.

Wilson, E. O. (1971). *The insect societies*. Cambridge, Mass.: Belknap/Harvard.

Section IV

Recent Research Issues

The Problem of Measuring Associative Learning in *Drosophila melanogaster* with Proper Procedures to Control for Nonassociative Effects

Michael G. Terry and Jerry Hirsch

Promise

Interest in *Drosophila melanogaster* as a model system for studying associative learning has increased dramatically in recent years (e.g., Stipp, 1991), because the well-understood genetic system of this species makes it unique among the metazoa for use in analysis of the genetic and other biological correlates of relatively simple learning processes. As they have characterized more elements of the behavioral repertoire and evaluated more behavioral effects of stimuli in *Drosophila*, drosophilists have been able to develop apparatuses and procedures in which they can expose flies to associations between stimuli and then measure behavioral changes which may indicate that the subjects are responding to the association. Evidence that the observed behavioral changes are attributable to associative learning rather than to nonassociative processes continues to improve as more appropriate control procedures are incorporated into *Drosophila* learning experiments.

Problem

Despite the historical trend of more appropriate controls being incorporated into more recent *Drosophila* learning experiments, most (if not all) published experiments still seem to be missing controls for one or more currently identifiable, nonassociative factors.[1] In the absence of an adequate set of control procedures, parsimony favors the (more simple) nonassociative factors over the (less simple) associative factors as the more likely explanations of the behavioral changes observed in these experiments. The identification of this problem (and attempts to solve it) has prompted vigorous debates (e.g., McGuire, 1984; Tully, 1984, 1986; Holliday & Hirsch, 1986a; Ricker, Hirsch, Holliday & Vargo, 1986; Dudai, 1988; Hirsch & Holliday, 1988), many of which have been clouded by confusions of terminology. We would like to further that debate and make our use of "learning terminology" explicit at the outset. We understand "associative learning" to be a biopsychological process inferred from a significant increase or decrease (from some baseline) in a subject's emission of a particular behavior, under an experimentally determined stimulus, that is attributable exclusively to the experimenter's association of that stimulus with another stimulus. We define a "stimulus" as an experimentally measured or manipulated event or set of events (usually of brief duration) that is perceptible to the subject (as demonstrated psychophysically) and note that such events may include observable elements of the subject's own activity in some experimental situations. We designate the "experimental association of stimuli" as a presentation of one stimulus (sometimes through defined temporal relationships to other "mediating" stimuli) where the temporal relationships usually are chosen in a *pragmatic* fashion to maximize the change in the measured behavior, within the practical constraints of a properly controlled associative-learning experiment. To us, the crucial control procedure in an associative-learning experiment—what we have started to call the "experimental dissociation of stimuli" (Rescorla, 1967)—is a presentation of two stimuli *without* a defined temporal relationship to one another (or without defined temporal relationships to other "mediating" stimuli) and which is *not* accompanied by the behavioral change (from some baseline) observed when the two stimuli are experimentally associated. In summary, our understanding is that associative learning is measured as a significantly greater increase or decrease in a subject's emission of a behavior under a stimulus when that stimulus is experimentally associated with another stimulus than when the two stimuli are experimentally dissociated. It is our hope that such an understanding may facilitate both the recognition of uncontrolled nonassociative factors and the development of more appropriate sets of control procedures, in a number of experimental paradigms, so that more conclusive evidence of associative learning in *D. melanogaster* may emerge.

Three Paradigms

Much of the work on biological bases of associative leaning in *D. melanogaster* has been focused on three paradigms: (1) appetitive conditioning of the proboscis extension reflex (PER) and bidirectional selective breeding to produce good and poor conditioning lines (here represented by Holliday & Hirsch, 1986b; Lofdahl, Holliday & Hirsch, 1992); (2) operant conditioning of flight orientation (here represented by Wolf & Heisenberg, 1991); and (3) aversive conditioning of odor avoidance (here represented by Tully & Quinn, 1985). A discussion of these three exemplars, though by no means a thorough review of the field, may illustrate how researchers have strived to develop stimulus presentation procedures, behavioral observation methods and experimental controls in an effort to measure associative learning while excluding nonassociative effects.

Holliday and Hirsch (1986b) measured the increased exposure to a salt solution for 5 seconds if that stimulus has been experimentally associated with a subsequent tarsal exposure to a sucrose solution for 5 seconds (available to the proboscis for 2–3 seconds), where the sucrose nonassociatively elicits a proboscis extension. It was discovered early on, however, that the sugar stimulation, used as the unconditional stimulus to elicit the PER, induces a US-sensitization or central excitatory state (CES) in flies (Dethier, Solomon & Turner, 1965; Vargo & Hirsch, 1982, 1985; also in honeybees, Brandes, Frisch & Menzel, 1988, p. 984), where CES has been defined as the increased probability that a food-deprived but water-satiated fly will extend its proboscis to a water stimulus applied to its chemoreceptors when that water stimulus is preceded by sucrose stimulation. Without proper CES controls, such a sensitization effect is easily misinterpreted as conditioning. Therefore, it was necessary to gain some understanding first of CES in *Drosophila* before attempting conditioning, as it had previously been necessary to do with the black blow fly *Phormia regina* (Dethier et al., 1965; Nelson, 1971). In fact, the *Drosophila* studies benefitted greatly from our experience with *Phormia* (Holliday & Hirsch, 1986b; Ricker, Hirsch, Holliday & Vargo, 1986).

The measurement of CES involved a water pretest, followed immediately by sucrose stimulation, followed by an interstimulus interval (ISI) and then a water posttest (Figure 1).

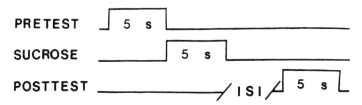

Figure 1. Stimulus schedule for one trial of a central excitatory state (CES) test. (Both pretest and posttest consisted of distilled water. ISI = interstimulus interval. From Vargo & Hirsch, 1982.)

In a water-satiated fly, the difference in response magnitude between the pretest and the posttest is the measure of the sucrose-induced CES, as shown in Table 1 from the experiment defining CES in *Drosophila* (Vargo & Hirsch, 1982).

Table 1. Comparison of Mean Scores of Stimuli for Each Group of Experiment 4

Group	n	Pretest	Sucrose	Posttest
1	31	3.452	12.00	7.226
2	30	3.333	—	3.367
3	30	—	12.00	8.100
4	31	8.097	12.00	—
5	31	4.226	—	—
6	31	—	12.00	—

(from Vargo & Hirsch, 1982)

Group 1 received the full test procedure: pretest, sucrose and posttest. The controls for that first group's defining results appear in the other 5 groups, from which one or two of the stimuli in the sequence have been omitted. The control results confirm the interpretation of the Group 1 results: Water produces a low level of responding (Groups 2 and 5), sucrose produces the highest level (Groups 1,3, 4 and 6) and it also increases the level of response to the first post-sucrose water stimulus (Groups 3 and 4) but not to a second one (Group 1), i.e., the first post-sucrose water stimulus discharges the measurable CES, which otherwise does not decrease over the 6–minute ITI (compare Groups 1, 3 and 4).

The CES measurements show that sucrose presentation induces a CES effect that must be taken into account on every conditioning trial.

In traditional terms, Holliday and Hirsch designated the salt solution as a conditioned (conditional) stimulus (CS), the proboscis extension to the salt solution as a conditioned (conditional) response (CR), the sucrose solution as an unconditioned (unconditional) stimulus (US) and the proboscis extension to the sucrose solution as an unconditioned (unconditional) response (UR) in a respondent (classical, Pavlovian) conditioning paradigm. In addition to doing control experiments assessing the extent of US-, and of CS-, sensitization independently of the conditioning procedure, they attempted to eliminate a nonassociative increase in the tendency to extend the proboscis to a variety of solutions after the fly has contacted sucrose with the labellum of its proboscis for 2–3 seconds (i.e., US-sensitization)[2], they inserted a 5–second intertrial stimulus (ITS) of tarsal

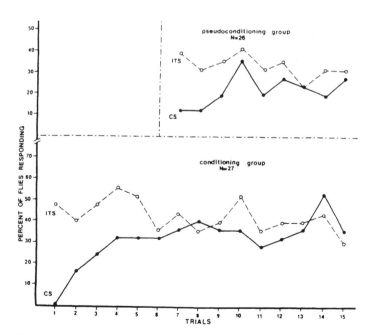

Figure 2. Percentage of animals responding to the conditioned stimulus (CS) and the intertrial stimulus (ITS) during conditioning (bottom) and sensitization induced by the unconditioned stimulus (US, top) (from Holliday & Hirsch 1986a).

stimulation with distilled water 170 seconds after the sucrose and 175 seconds before the salt. They also used measurements of intertrial responses (ITRs) to the ITS as indicators of US-sensitization and contrasted the nonsystematic fluctuations of ITRs with the systematic changes in CRs across trials (see Figure 2) to show that the effect of US-sensitization was distinguishable from the associative effect of "CS-US pairings."

In addition, Holliday and Hirsch (1986b) made "unpaired presentations of the CS and US" (by moving the former from approximately 0.5 seconds to 90 seconds before the latter) and showed that the CRs became much less frequent while ITRs remained just as frequent (see Figure 3), suggesting that the magnitude of the associative effect was influenced by the longer CS-US interval but not that of the nonassociative CES effect.

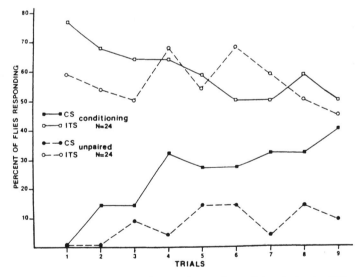

Figure 3. Percentage of animals responding to the conditioned stimulus (CS) and the intertrial stimulus (ITS) during conditioning and unpaired procedures (from Holliday & Hirsch 1986a).

The evidence for conditioning *Drosophila* was obtained with an unselected (wild-type) laboratory population, in which only a minority of the animals (about 24%) qualified as having been successfully conditioned; i.e., as shown in Figure 4, there were reliable individual differences (IDS) in response to the conditioning procedures.

From the distribution of IDs in acquisition scores in the foundation population, extreme scoring individuals were chosen for bidirectional selective breeding to produce good and poor conditioning lines, as had previously been done with *P. regina* (McGuire & Hirsch, 1977). The first attempts to breed *Drosophila* selectively for

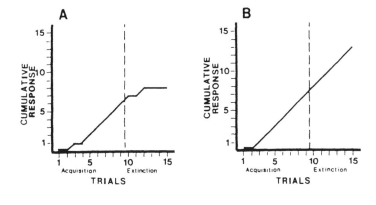

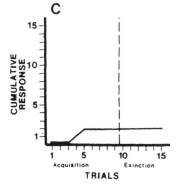

Figure 4. Cumulative responses to the conditioned stimulus (CS) over 9 trials of conditioning and 6 trials of unpaired extinction for (1) an individual representing those flies (8 out of 34, 24%) showing both conditioning and extinction, (2) an individual representing those flies (5 out of 34, 15%) showing acquisition without extinction and (3) an individual representing those flies (12 out of 34, 35%) responding but showing neither conditioning nor extinction (not represented is a fourth type showing total absence of responding to the CS) (from Holiday & Hirsch, 1986a).

conditionability produced excitable, but few conditionable, flies. Reexamination of the basic evidence for conditioning, in conjunction with which we had also included extinction trials (Holliday & Hirsch, 1986b), revealed two types among the individuals showing good acquisition—those that extinguished and those that failed to show unpaired extinction. We realized that excitable flies would continue to respond in unpaired extinction training, during which time conditioned flies do extinguish. Since acquisition scores alone do not separate the two types, inadvertently we had probably bred to increase the proportion responding because of US-sensitization (CES). Therefore, to the acquisition criterion we added an extinction criterion for selection, and this time we succeeded in breeding flies for conditionability that were not merely more excitable. The percentage of animals showing good conditioning increased from 19% in the foundation population to 77% over 25 generations of selection, but the level of their CES response did not increase (see Figure 5).

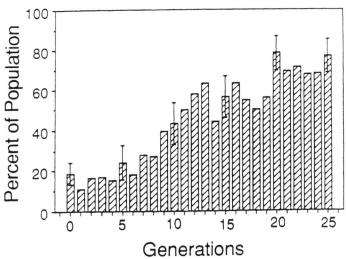

Figure 5. Percentage of subjects showing both an acquisition and extinction pattern from the good conditioning population over generations of selective breeding (from Lofdahl, Holliday & Hirsch, 1992).

Furthermore, for the first time the level of the acquisition response in the conditioned flies climbed above that of their CES response (see Figure 6), i.e., to a level comparable to that reported for honey-

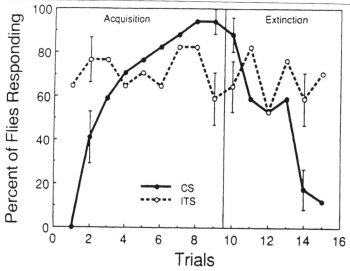

Figure 6. Response levels to the conditioned stimulus (CS) and the intertrial stimulus (ITS) for the 18 (of 23) flies meeting the good conditioning criteria (including extinction) from Generation 20 of the good conditioning population (from Lofdahl, Holiday & Hirsch, 1992).

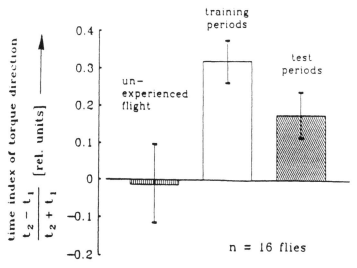

Figure 7. Time index of torque direction (where positive indices indicate torquing away from the direction associated with heat) during unexperienced flight (baseline), training period and test period (from Wolf & Heisenberg, 1991). Reprinted by permission of Springer-Verlag New York, Inc.

bees (in the bee research, unfortunately, the ITS control for CES has not been employed on each conditioning trial [Lofdahl, Holliday & Hirsch, 1992]).

Wolf and Heisenberg (1991) tethered a fly to a wire, allowed the fly to beat its wings so that it torqued to its left or right, associated torquing to one half-space (e.g., left of center) with exposure to intense heat and observed a decrease (compared to baseline) in the amount of time that the fly torqued to that side, even after the heat was switched off (see Figure 7).

In traditional terms, we can designate one half-space (e.g., the fly's left) as the operant occasion (O, or discriminative stimulus), the other half-space (e.g., the fly's right) as the operant nonoccasion (~O, or nondiscriminative stimulus), the fly's torquing toward O (e.g., torquing left) as the operant behavior (B), the fly's not torquing toward O (e.g., not torquing left) as the operant nonbehavior (~B), presentation of the heat as the operant consequence (C) and omission of heat as the operant nonconsequence (~C) in an operant (instrumental, Skinnerian) conditioning paradigm. The direction of torquing associated with heat was counterbalanced across subjects to control for individual differences in the (nonassociative) tendency to torque left or right.

Tully and Quinn (1985) modified an operant conditioning apparatus and procedures developed by Quinn, Harris and Benzer (1974) into a respondent conditioning paradigm by placing groups of $\approx$ 150 flies in a tube lined with an electrifiable grid, exposing them to one odorant ("A") for 60 seconds in association with the intermittent electrification of the grid, exposing them to unscented air for 30 seconds while not electrifying the grid, exposing them to another odorant ("B") for 60 seconds while not electrifying the grid, transferring them to a choice point between a tube containing "A" and a tube containing "B" and observing that fewer flies moved into the "A" tube than the "B" tube over the course of 120 seconds. In traditional terms, they designated "A" as a positive conditioned (conditional) stimulus (CS+), "B" as a negative conditioned (conditional) stimulus (CS-), the tendency of most flies to move away from "A" rather than away from "B" a conditioned (conditional) "avoidance" response (CR), electrification of the grid as an aversive unconditioned (unconditional) stimulus (US) and the tendency of most flies to move away from the electrified grid (measured in a separate experiment) as an unconditioned (unconditional) "escape" response (UR) in a respondent (classical, Pavlovian) conditioning paradigm. The concentrations of 3–octanol (OCT) and 4-methylcyclohexanol (MCH) were adjusted so that the tendency of test flies to walk away from OCT was approximately equal to their tendency to walk away from MCH. Designation of OCT and MCH as odorants "A" and "B" was counterbalanced across groups of flies to control for such compound-specific nonassociative effects. When the order and intervals between the CS+, CS- and US were varied, a majority of the flies moved away from the CS+ only if it had been presented simultaneously with or shortly after the US (see Figure 8), suggesting that the flies were responding to the association of the CS+ with the US.

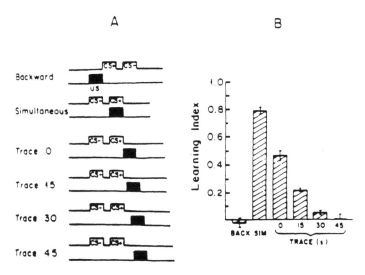

Figure 8 A, B. Temporal specificity of CS-US pairings. (A) Stimulus schedules for backward conditioning, simultaneous conditioning and four trace conditioning intervals. Odor presentations (CS+ or CS-) lasted for 60 seconds; the length of the baselines indicates the amount of time flies spent in the training tube. (B) Mean learning indices for these CS-US pairings indicate that the US must follow CS presentations closely in order to produce strong learning, but they suggest the persistence of a CS trace. Each point represents 6–8 experiments (from Tully & Quinn 1985). Reprinted by permission of Springer-Verlag New York, Inc.

Six Uncontrolled Factors

In the three previously described *Drosophila* learning paradigms, despite their excellence in comparison to most other paradigms in the field, there are occasional gaps in control procedures related to at least six different factors: (1) independent determination of subject responsiveness to *each* of the experimental stimuli; (2) confirmation of the experimental *independence* of individual behavioral observations for each of many subjects that are run together as a group or for each of many intervals that are run by a single subject; (3) distinction of responses to *single* dissociated stimuli from responses to the experimental association of two stimuli; (4) exclusion of response *decrements* stemming from repeated exposure to *single* controlled or uncontrolled stimuli (i.e., habituation); (5) exclusion of response *increments* stemming from exposure to any *one* of a number of controlled or uncontrolled stimuli (i.e., sensitization); and (6) exclusion of a *reversal* in a response *decrement* from the repeated exposure to *one* stimulus either by subsequent exposure to another controlled or uncontrolled stimulus or by the passage of time (i.e., dishabituation). A description of how these factors may affect behavioral observations when insufficient controls are in place may serve to illustrate the problem of measuring associative learning in such a way that nonassociative effects can be excluded experimentally or analytically.

In order for a subject to perceive an experimental association between two stimuli, it must actually perceive *each* of the two stimuli as they are presented in the experimental situation. If the subject actually perceives only one of the stimuli, then the attribution of a change in its behavior to associative learning is necessarily mistaken. The lack of an independent determination of subject responsiveness to each of the experimental stimuli is primarily a problem in *Drosophila* learning experiments where the flies are run as groups. In this situation, it is practically impossible to assure or measure that any particular subject is exposed to the two stimuli in such a way that both stimuli are actually perceptible. Most experiments of this type are designed to minimize this problem by presenting stimuli that may envelop all of the flies continuously, such as widespread odorants or diffuse lights. Although it is often debatable whether such forms of stimulation affect every fly in the group (particularly when they are interacting with one another during the experiment), our primary concern is with types of stimulus presentation that have been shown to fail to affect at least some flies some of the time. For example, in experiments where flies are free to move about on an

electrified grid (Tully & Quinn, 1985), it is *not* safe to assume that
any particular fly is actually stimulated electrically. With the grids
used in such experiments, one of the two conductors and the insulator must (together) cover more than half of the total surface area, so
it is not unlikely that a fly would find itself on a single conductor
and/or on an insulator and thus "out of the circuit." Although we do
not have the space to go into the details, we can note that several
researchers have observed free-moving individual flies on various electrified grids in our laboratory, and they have all confirmed that many
flies may spend most or all of their time walking or standing on a
single conductor and/or on an insulator so that electrical current is
not flowing through the fly. (As an aside, whether or not flies can
learn an association between touching oppositely charged conductors and receiving electrical current is an open [and experimentally
difficult] question.) As this example suggests, demonstrations of effective presentation of experimental stimuli are much easier when
the stimuli are presented to individual subjects. However, such demonstrations are useful only if they are used to screen out subjects that
do not perceive both stimuli during the experimental association of
stimuli. In the proboscis-extension conditioning paradigm (Holliday
& Hirsch, 1986b), unfortunately, many subjects did not actually
come into contact with one or more of the stimuli on some trials
because their tarsi were not extended. Although the 15 out of 233
(6.4%) of the subjects that did not have a tarsal exposure to any one
of the stimuli on more than 33% of presentations were dropped
from the experiment (without replacement), some of the remaining
subjects did *miss* some stimulus exposures. This is a problem because there is no analytical distinction between other associative or
nonassociative effects and: (1) failure to respond to the CS because
the CS was not actually presented to the tarsi on *that* trial; (2) failure
to respond to the CS because the CS was not actually presented to
the tarsi soon before the US on one or more *preceding* trials; (3)
failure to respond to the CS because the US was not actually presented to the tarsi soon after the CS on one or more *preceding* trials;
(4) failure to respond to the ITS because the *preceding* US was not
actually presented to the tarsi; or (5) response to the CS because the
preceding ITS was not actually presented to the tarsi after exposure
to the sensitizing US. In addition, in subjects that missed some US

presentations to the labellum of the proboscis, there is no analytical distinction between other associative or nonassociative effects and: (1) failure to respond to the CS because the *preceding* CS was not soon *followed* by labellar exposure to the US; (2) failure to respond to the CS because there was no *preceding* sensitizing labellar exposure to the US; or (3) failure to respond to the ITS because there was no *preceding* sensitizing labellar exposure to the US. Because the raw data on stimulus exposures and responses to stimuli are still available (Holliday, 1984), it may be possible to address these problems post hoc. In general, where the subjects' uncontrolled behavior may affect whether the experimental stimuli actually reach them, it does not seem safe to attribute any behavioral change to an association between two stimuli unless the experimenters have demonstrated that they have *effectively* presented *each* stimulus to each subject.

Whether the measurement of a change in behavior involves a single trial with many subjects or repeated trials with a single subject, pooling of the data is inappropriate if the individual observations are not experimentally independent of one another (as specified in the assumptions underlying the statistical tests applied to such data). For example, in experiments where flies are run together as a group (Tully & Quinn, 1985), it is *not* safe to assume that the behavior of each individual is independent of that of other individuals. In fact, there is ample evidence that flies in groups interact with one another by aggregating, dispersing and following (e.g., Walton, 1968; Hay & Crossley, 1977; Pluthero & Threlkeld, 1979). Indeed, the hypothesis that a count of the individuals in a group behaving in a particular way (e.g., ending up in the same location) reflects interactions between subjects rather than the effect of a commonly experienced experimental manipulation cannot be rejected without disconfirming results from an appropriately-powerful set of experiments designed to measure such interactions between flies under conditions closely resembling those in the experiment of interest. As another example, in experiments where single subjects are observed over successive intervals (e.g., Holliday & Hirsch, 1986b; Wolf & Heisenberg, 1991), the subject's behavior during one interval may resemble its behavior during the preceding interval even though the experimental conditions may have changed (e.g., shifting from "training" to an "extinction posttest"). Indeed, the hypoth-

esis of such a behavioral continuity cannot actually be rejected unless an appropriately powerful time-series analysis (e.g., autocorrelation) indicates that the component of the observed behaviors that cannot be predicted from the experimental manipulations
of stimuli (i.e., the residuals) also cannot be predicted from previously observed behaviors (i.e., preceding values of the dependent
variable or of its residuals). Therefore, data pooled across subjects
run as a group are not useful without appropriate controls for behavioral interactions between subjects, and data pooled across successive observations of a single subject are not useful without appropriate controls for behavioral continuities between observations. The
appropriate controls are implemented much more easily in the latter
case, so it seems that evidence of associative learning is more likely
to be obtained from experiments in which subjects are stimulated
and observed individually.

The simplest way to distinguish between a behavioral change in
response to an *experimental association* between stimuli and a behavioral change in response to either of the two stimuli taken alone is to
present the two stimuli in a *dissociated* manner and to observe that
the behavioral change observed under association is not observed
under dissociation. This distinction has not been made, for example,
in Wolf and Heisenberg's (1991) experiment because the consequen-
tial heat stimulus was always applied when the fly torqued in the
designated position. Because the fly's behavior was not measured
while the heat stimulus was presented in a dissociated manner with
respect to its torquing movement, it is possible to interpret the observed behavioral change in terms of the direct effects of the heat
stimulus alone. Specifically, if heat increased *random* torquing activity (i.e., torquing in *both* directions with *equal* probability) and the
lack of heat decreased such activity, then the torquing would tend
toward the position in which there was a lack of heat merely because
that position served as an "activity sink." In order to distinguish
between positive thermokinesis (increased activity under thermal
stimulation) and associative learning in such an experiment, each fly
could be observed to torque away from the designated direction in a
subsequent test with the heat presented *independently* of the direction of torque. Instead of this, Wolf and Heisenberg (1991) simply
disconnected the heat source and observed that flies continued to

torque away from the designated direction. Not only does this fail to control for any nonassociative effects of the heat *per se*, but also the time spent torquing away from the designated direction is not measured independently of the fly's immediately preceding behavior. Specifically, just before the heat was turned off, the fly tended to be torquing away from the designated direction and immediately after the heat was turned off the fly may have tended to continue torquing in the *same* direction.

The former tendency may be a (direct) nonassociative effect of the heat and the latter tendency may be a nonassociative effect of behavioral *continuity*. Randomized presentations of the consequence with respect to occasion-behavior conjunctions during a number of discrete trials would provide an adequate posttest in this experimental paradigm.

In experiments where the stimuli are presented repeatedly or prolongedly, it is crucial that the nonassociative effects of habituation be excluded with proper control procedures. This issue is most evident when the emission of the measured behavior decreases, but habituation of a competing behavior or of a behavioral "inhibition" may lead the emission of the measured behavior to increase. For example, Holliday and Hirsch (1986b) repeatedly presented a salt solution in their classical conditioning procedure and observed an increase in proboscis extension to the salty stimulus. When the salt solution was presented alone, it seemed to *inhibit* proboscis extension as compared to a distilled-water stimulus, but the experimenters were not in the position to measure habituation of this inhibition because the response level to the salty stimulus was essentially zero. In the conditioning procedure, however, sucrose presentations increased the tendency of the flies to extend their proboscises to a *variety* of solutions, and they did begin responding to the *salt* solution. The observed increase in responding to the salt solution could very well reflect habituation of the inhibition of responding to that stimulus, rather than an associative effect. One way to control for such a habituation effect would be to use two different ("inhibitory") salt solutions as CS+ and CS-, reversing their roles in different phases of the experiment (see Ricker, Brzorad & Hirsch, 1986) and to use varying interstimulus intervals to control for temporal factors, which play a strong role in habituation. Under such a control procedure, an increase in responding to one stimulus as its role changes from CS- to CS+ and a decrease in responding to the other stimulus as its role changes from CS+ to CS- would not be attributable to differences in habituation of responding to the two stimuli.

Presentation of a stimulus may nonassociatively increase responsiveness to that stimulus and/or to another stimulus through sensitization, which complicates the interpretation of associative-learning experiments where the subjects respond more to a stimulus when it and/or another stimulus is presented repeatedly or prolongedly. For example, Holliday and Hirsch (1986b) found that presentation of a sucrose solution nonassociatively increased the response of flies to other solutions. They attempted to eliminate this sensitization by interposing a water stimulus (ITS) between the sucrose US and salt solution "CS," but there is no conclusive evidence that the procedure was completely effective. Perhaps the insertion of multiple water stimuli in this experimental paradigm would permit the observation of a diminution of sucrose-induced sensitization to a zero level. More generally, one should note that sensitization effects are temporally variable and that parametric evaluation of such sensitization effects across a wide range of interstimulus intervals would go a long way toward resolving this issue.

In experiments where one or more stimuli are presented repeatedly or prolongedly, perhaps leading to habituation of an existing nonassociative response, the subsequent passage of time and/or exposure to another form of stimulation may (nonassociatively) dishabituate that response. For example, in the experiment of Tully and Quinn (1985), the CS+ is presented at a longer interval before the US than is the CS- so that the *longer dishabituation interval* for the CS+ could lead to a relatively greater *nonassociative* component of response to that stimulus (i.e., more walking away from the CS+ odorant). Furthermore, the delivery of a (potentially) *dishabituating* stimulus (i.e., electrical current) during prolonged presentation of the CS+ similarly could lead to a greater *nonassociative* effect of the CS+ on the measured behavior. Thus, the tendency to walk away from the CS+ and toward the CS- in the Tully and Quinn (1985) aversive conditioning paradigm may reflect an experimentally imposed relative difference in the dishabituation of the tendency to walk away from either odorant. In order to distinguish between dishabituation and a response to the experimental association of stimuli, generally it is simplest to control for habituation processes.

Because several nonassociative effects of the same type may be present simultaneously (e.g., habituation of different responses to the same stimulus and of the same or different responses to various stimuli) and because nonassociative effects of different types may interact with one another (e.g., dishabituation of a habituated response), the more control procedures that are lacking in an experiment, the more complicated and varied the potential interpretations of the results. Furthermore, to the extent that conditions differ between an "associative-learning experiment" and a *separate* "control experiment," additional *nonassociative* differences between experimental treatments arise, making it difficult to attribute a behavioral difference between experiments to *associative* factors. Rather than continuing to dwell on such *problems*, however, we would prefer to proffer a potential *solution*.

One Potential Solution

To show that at least one solution to the problem of measuring associative learning in *D. melanogaster* may be attainable in the near term, we would like to provide a brief and generic example of an operant conditioning procedure that we are planning to apply in our laboratory. Through screening of unselected populations and/or selective breeding, we can obtain flies with various combinations of tendencies to walk toward or away from earth or light. For example, we might obtain a fly that tends to walk more away from earth (upward) than toward earth (downward) but equally toward light (lightward) or away from light (darkward). We can expose such a fly to a five–step operant-conditioning procedure designed to evaluate whether or not presentation of an opportunity to walk upward (i.e., an opportunity to engage in a high-probability or "preferred" behavior) positively reinforces walking either lightward or darkward (i.e., expression of a low-probability or "nonpreferred" behavior). In the first step, the fly is exposed to 24 discrete trials in which it can walk either lightward or darkward, during a brief interval (4, 8, 12, or 16 seconds, where this duration is randomized across trials), with a subsequent opportunity to walk upward (for 10 seconds) on half of those trials (in a randomized sequence, independent of the fly's behavior). In the second step, the fly is presented with the opportunity to walk upward (the operant consequence, C_2, where the subscript refers to the step number) during each of 24 trials only if and when it walks at least 1 cm (the operant behavior, B_2) *lightward* (the operant occasion, O_2) during the designated interval. In the third step, the fly is exposed to another 24 trials in which it can walk lightward or darkward with a subsequent opportunity to walk upward on half of those trials. In the fourth step, the fly is presented with the opportunity to walk upward (C_4) during each of 24 trials only if and when it walks at least 1 cm (B_4) *darkward* (O_4) during the designated interval. In the fifth step, the fly is exposed to still another 24 trials in which it can walk lightward or darkward with a subsequent opportunity to walk upward on half of those trials. Note this five-step operant-conditioning procedure can be adapted to the different behavioral tendencies of other flies (e.g., one that walks more lightward than darkward but equally upward and downward) by substituting the requirement of walking in any two (opposite) low-probability directions (e.g., upward in step 2 and downward in step 4) for walking lightward or darkward (respectively) and by substituting the opportunity to walk in any high-probability direction (e.g., lightward) for walking upward.

This experimental design has several features that, collectively, control for the previously described nonassociative factors that may confound the measurement of associative learning in the three paradigms discussed above. Measurements of the tendencies to walk toward or away from earth or light during steps 1, 3 and 5 provide an independent determination of each subject's responsiveness to each of the relevant stimuli. Time-series analysis (e.g., autocorrelation) can test the experimental independence of each fly's successive responses to light during each of the five steps. Parametric and randomized variation of trial durations during each of the five steps allows for the assessment and analytic control of interstimulus-interval effects, which may act either directly or through differences in habituation, sensitization or dishabituation processes. Because the opportunity to walk upward is not presented until the fly has walked at least 1 cm lightward (in step 2) or darkward (in step 4), the fly's *preceding* response to the light is observable separately from its *subsequent* response to the opportunity to walk upward. Reversal of the experimental association of stimuli between steps 2 and 4 controls for any *fixed* tendency to walk more lightward or darkward. In addition, the reversal (i.e., $O_2 = {\sim}O_4$, $O_4 = {\sim}O_2$, $B_2 = B_4$, and $C_2 = C_4$) allows for analytical control of any behavioral *shift* (in the tendency to walk lightward or darkward) attributable to habituation, sensitization or dishabituation differences (e.g., related to differences in interstimulus intervals) between steps 2 and 4. Habituation, sensitization and dishabituation can also be evaluated across the entire experimental session by analyzing any consistent trend in the fly's tendency to walk more lightward or darkward from step 1 through step 3 to step 5. After using these analyses to estimate the fly's expected responses to light in steps 2 and 4, a residual tendency to walk more *lightward* during step 2 and more *darkward* during step 4 could only be attributed to an associative effect.

Perspective

We have focused on the *problem* of measuring associative learning in *D. melanogaster* with proper procedures to control for nonassociative effects. It is important that we realize that *there is such a problem* and that we recognize the *absence* of some appropriate control procedures in most if not all published reports in the field. As a result of this problem, some of the reported behavioral measures of associative learning might be explained more parsimoniously as measures of various nonassociative effects. On the other hand, the reliable and valid measurement of any particular nonassociative effect requires its own set of control procedures—because most *Drosophila* learning experiments do not happen to include such controls, it is not safe to attribute any of their measured changes in behavior to any single nonassociative effect. Rather than attempting to analyze all of the nonassociative effects that may be present in inadequately controlled associative-learning experiments in an effort to tease out residual associative effects, we propose that an adequate set of appropriate control procedures be incorporated into the design of each *Drosophila* learning experiment so that associative learning can be measured as directly as possible. This strategy should not only yield more reliable and valid measures of phenotypes related to associative learning but also simplify the incorporation of appropriate control procedures for measuring genetic and other biological correlates of individual differences in such phenotypic measures. If we can bring an adequate set of control procedures to bear in a single experiment (and in another and another), then we may finally begin to realize the enormous potential of *D. melanogaster* as a model system for evaluating the biological bases of associative learning.

Notes

1. This problem, and the complications that it brings, is by no means unique to learning research with *Drosophila melanogaster* in particular or to learning research with invertebrates in general. It is our opinion that learning research with *any organism* could benefit from the sort of intensive evaluation and improvement of experimental designs that is presented here.
2. Note that this form of US-sensitization in flies is often referred to as the central excitatory state (CES), following the work of Dethier and others (Dethier, et al. 1965; Vargo & Hirsch 1982, 1985).

References

Brandes, C., Frisch, B., & Menzel, R. (1988). Time course of memory formation differs in honey bee lines selected for good and poor learning. *Animal Behaviour, 36*, 981–985.

Dethier, V. G., Solomon, R. L., & Turner, L. H. (1965). Sensory input and central excitation and inhibition in the blowfly. *Journal of Comparative and Physiological Psychology, 60*, 303–313.

Dudai, Y. (1988). Neurogenetic dissection of learning and short-term memory in *Drosophila. Annual Review of Neuroscience, 11*, 537–563.

Hay, D. A., & Crossley, S. A. (1977). The design of mazes to study *Drosophila* behavior. *Behavior Genetics, 7*, 389–402.

Hirsch, J., & Holliday, M. (1988). A fundamental distinction in the analysis and interpretation of behavior. *Journal of Comparative Psychology, 102*, 372–377.

Holliday, M. (August 1984). Classical conditioning of the proboscis extension reflex in *Drosophila melanogaster*. Unpublished Master's thesis, University of Illinois Urbana-Champaign.

Holliday, M., & Hirsch, J. (1986a). Excitatory conditioning of individual *Drosophila melanogaster. Journal of Experimental Psychology: Animal Behavior Processes, 12*, 131–142.

Holliday, M., & Hirsch, J. (1986b). A comment on the evidence for learning in Diptera. *Behavior Genetics, 16*, 439–447.

Lofdahl, K., Holliday, M., & Hirsch, J. (1992). Selection for conditionability in *Drosophila melanogaster. Journal of Comparative Psychology, 106*, 172–183.

McGuire, T. R. (1984). Learning in three species of Diptera: The blow fly *Phormia regina*, the fruit fly *Drosophila melanogaster*, and the house fly *Musca domestica. Behavior Genetics, 14*, 479–526.

McGuire, T. R., & Hirsch, J. (1977). Behavior-genetic analysis or *Phormia regina*: Conditioning, reliable individual differences, and selection. *Proceedings of the National Academy of Sciences USA, 74*, 5193–5197.

Nelson, M. (1971). Classical conditioning in the blowfly (*Phormia Regina*): Associative and excitatory factors. *Journal of Comparative and Physiological Psychology, 77*, 353–368.

Pluthero, F. G., & Threlkeld, S. F. H. (1979). Some aspects of maze behavior in *Drosophila melanogaster. Behavioral and Neural Biology, 26*, 254–257.

Quinn, W. G., Harris, W. A., & Benzer, S. (1974). Conditioned behavior in *Drosophila melanogaster. Proceedings of the National Academy of Science USA, 71*, 708–712.

Rescorla, R. A. (1967). Pavlovian conditioning and its proper control procedures. *Psychological Review, 74*, 71–80.

Ricker, J., Brzorad, J., & Hirsch, J. (1986). A demonstration of discriminative conditioning in the blow fly, *Phormia regina. Bulletin of the Psychonomic Society, 24*, 240–243.

Ricker, J. P., Hirsch, J., Holliday, M. J., & Vargo, M. A. (1986). An examination of claims for classical conditioning as a phenotype in the genetic analysis of diptera. In J. L. Fuller & E. C. Simmel (eds.), *Perspectives in Behavior Genetics* (pp. 155–200). Hillsdale, N.J.: Lawrence Erlbaum Associates.

Stipp, D. (1991). Grand Gamble on Fruit Fly Learning: Cold Spring Harbor is betting millions that the genetic workhouse *Drosophila* can provide a good model system for studying memory, but many in the field remain skeptical. *Science, 253*, 1486–1487.

Tully, T. (1984). *Drosophila* learning: Behavior and Biochemistry. *Behavior Genetics, 14*, 527–557.

Tully, T. (1986). Measuring learning in individual flies is not necessary to study the effects of single-gene mutations in *Drosophila*: A reply to Holliday and Hirsch. *Behavior Genetics, 16*, 449–455.

Tully, T., & Quinn, W. G. (1985). Classical conditioning and retention in normal and mutant *Drosophila melanogaster. Journal of Comparative Physiology, 157*, 263–277.

Vargo, M., & Hirsch, J. (1982). Central excitation in the fruit fly (*Drosophila melanogaster*). *Journal of Comparative and Physiological Psychology, 96*, 452–459.

Vargo, M., and Hirsch, J. (1985). Behavioral assessment of lines of *Drosophila melanogaster* selected for central excitation. *Behavioral Neuroscience, 99*, 323–332.

Walton P. D. (1968). Factors affecting geotaxis scores in *Drosophila melanogaster, Journal of Comparative Physiological Psychology, 65*, 186–190.

Wolf, R., & Heisenberg, M. (1991). Basic organization of operant behavior as revealed in *Drosophila* flight orientation. *Journal of Comparative Physiology, 169*, 699–705.

Pseudostinging Behavior of the Harvester Ant *Messor sanctus* Forel as a Response to Novelty

Ewa Joanna Godzińska, Anna Szczuka and Julita Korczyńska

Ants that regularly include seeds in their diet are called harvester ants. Harvester ants are not a monophyletic group: feeding on seeds is found in numerous species of the subfamilies Ponerinae, Myrmicinae and Formicinae (Hölldobler & Wilson, 1990). The ants of the genus *Messor* (subfamily Myrmicinae) belong to the best-known harvester ants. They are mainly granivorous, but their diet may include also dead and living insects and other proteinic food, such as snails and pieces of flesh of dead mammals (Cerda & Retana, 1994; Gillon, Adam & Hubert, 1984; Hahn & Maschwitz, 1985; López, Serrano & Acosta, 1992; Sheata & Kaschef, 1971).

As many highly evolved ants, the ants of the genus *Messor* have lost a functional sting in the course of their phylogenesis: their sting is fully atrophied (Maschwitz, 1975). We report here the results of field and laboratory observations and experiments demonstrating that these ants may, nevertheless, display stinging-like behavior, called by us "pseudostinging behavior." We observed, namely, that excited workers of *M. sanctus* often seize various objects that are close to them and then touch them with the gaster tip in a stinging-like manner. A response involving single or repeated touching of the object with the gaster tip was called by us the "true pseudostinging response." Sometimes the ant only flexes its gaster in the direction of the object seized by it; such behavior was called by us "gaster flexing." As used in this paper, "pseudostinging behavior" includes both gaster flexion and true pseudostinging responses.

Experiments

Field Observations of Pseudostinging Behavior of *Messor sanctus*

Our research on the causation and function of the pseudostinging behavior of the ants of the genus *Messor* started from field observations made by Godzińska and Lenoir in southern France (Leucate near Perpignan) in May 1991. In a pilot experiment on the orientation of the foragers of *Messor sanctus* Forel, a small stone was laid across a trail of a large colony of that species to investigate the responses to that obstacle in the outgoing and ingoing foragers. The ants responded to that stone by increasing the speed and sinuosity of their locomotion; as a consequence they aggregated at both sides of the stone. Many foragers carrying loads dropped them. However, as a rule these loads were not abandoned; ultimately many of them were picked up again either by the workers that had dropped them or by their nestmates aggregating nearby. Sometimes the ant that had detected such an abandoned seed simply picked it up in its mandibles and carried it away. However, frequently the detection of an abandoned seed triggered a more complex behavioral sequence in which seizing of the seed was followed by gaster flexing in its direction, and sometimes also by a single or repeated stinging-like touching of the seed with the gaster tip (Figure 1).

The exact behavioral pattern triggered in response to abandoned seeds was quite variable. Sometimes the ant performed only a rapid forward gaster flexing in the direction of the seed, without touching it with the gaster tip (Figure 1A). However, often the ant touched the seed with its gaster tip in a stinging-like manner (Figure 1B, C). In some cases the ant drew the tip of its gaster briefly against the surface of the seed in a sinuous way. That behavior was suggestive of both stinging behavior and chemical marking of the seed with the secretion of some abdominal gland(s), and we called it "pseudostinging behavior." The responses involving touching of the seed with the gaster tip were labeled by us as "true pseudostinging behavior," and those which consisted only in flexing of the gaster in the direction of the seed were labeled "gaster flexing."

Pseudostinging responses were followed either by the transport of the seed or by its abandonment. In some cases, after an act of gaster flexing, the ant remained immobile with the gaster curled

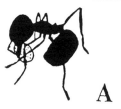

Figure 1. Pseudostinging behavior of workers of Messor sanctus *Forel. A: gaster flexing in the direction of a seed. B and C: true pseudostinging. Drawings based on the photographs taken by A. Lenoir in the field in Leucate (southern France).*

forward between the legs for several seconds. Pseudostinging behavior was carried out by workers belonging to all size categories, including very small ones.

The analysis of the photographs taken in the field in Leucate during our field observations of the pseudostinging behavior of *M. sanctus* revealed that the ants directed their pseudostinging behavior to objects held by them in the mandibles, often also by their forelegs and sometimes also by their middle legs (Figure 1). Pseudostinging ants often adopted an upright posture, with the thorax held almost vertically (Figure 1B, C). However, they were also observed to perform pseudostinging in a normal walking position, with the thorax held horizontally (Figure 1A).

During the observations of the pseudostinging behavior of *M. sanctus* carried out in the field in Leucate, a question arose as to whether these ants display pseudostinging behavior only in response to seeds that have been dropped by their nestmates. To answer that question, we collected similar seeds in the vicinity of the trail and placed them on the trail near the stone laid across it. The ants responded to these seeds in the same way as to seeds dropped by their nestmates. They also showed pseudostinging behavior in response to other varieties of seeds and to other fragments of vegetal matter placed by us on their trail close to the stone obstructing their way.

To throw more light on the causal factors and the function of pseudostinging behavior of *M. sanctus*, we decided to study it in the laboratory.

Laboratory Research on the Proximate Causal Factors of Pseudostinging Behavior of *M. sanctus*

Animals and Maintenance. A fragment of a large colony of that species containing a queen, about 2,000 workers and abundant brood was collected in Leucate immediately after the field observations described above. The ants were taken to our laboratory, first in Villetaneuse (France), then in Warsaw (Poland). In July 1991, the colony was composed of a queen, 1711 workers and abundant brood (eggs, larvae and pupae). It was housed in seminatural conditions in a relatively large (70 cm ¥ 55 cm ¥ 10 cm) flat tub filled with about a 5-cm layer of sand (Figure 2). The walls of the tub were painted with Fluon (PTFE), a substance used in myrmecological research to prevent the ants from escaping from artificial nests. The ants were allowed to settle in several large test tubes filled to one-third of their length with water trapped in by cotton plugs. To assure darkness, the tubes were wrapped in aluminium foil. The surface of the sand filling the tub served as the foraging area. The ants were fed on mixed birdseed, bread crumbs and crushed oatflakes. We also provided honey mixed with crushed apples and sand to make the mixture less sticky.

As noted in the literature, humidity has a strong stimulating effect on the foraging activity of the harvester ants (Gillon, Adam & Hubert, 1984; Johnson, 1991). Hence, every morning we sprayed the surface of the sand with water. Under these conditions, the ants showed a fairly high level of outside activity, in particular during the hours following the morning "dew." Moreover, although they continued to use the test tubes with water held in by cotton plugs as their brood chambers, they excavated other nest chambers in the moist sand at the opposite side of the tub, near the place where the

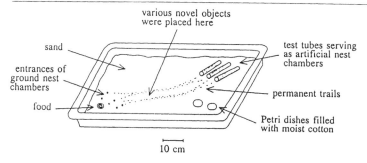

various novel objects
were placed here

sand

test tubes serving
as artificial nest
chambers

entrances of
ground nest
chambers

permanent trails

food

Petri dishes filled
with moist cotton

10 cm

Figure 2. Artificial nest used to house the captive colony of M. sanctus *used in our experiments.*

food was offered to them (Figure 2). The queen and the brood could be found either in one of the test tubes or in one of the ground chambers. When moving from the tubes to the entrances of their ground nests and to food, the ants as a rule followed two nearly parallel permanent trails (Figure 2).

Experiment 1, Series 1–4

Procedure. During the winter 1991/1992, our colony of *M. sanctus* was used in a series of tests carried out to investigate more closely the proximate mechanisms underlying pseudostinging behavior, true pseudostinging responses and gaster flexion of these ants. In particular, we asked the following questions:

1. What objects may trigger pseudostinging behavior in *M. sanctus?*
2. Do *M. sanctus* show greater readiness to display pseudostinging behavior in response to insect prey than in response to other food items and/or inedible objects?
3. Does the presence of insect prey modulate (enhance or suppress) the frequency of pseudostinging behavior directed by *M. sanctus* toward other objects offered to them simultaneously with that prey? In other words, do the same stimuli trigger both pseudostinging behavior oriented toward their source and modulate the frequency of pseudostinging responses oriented toward other objects?

To answer these questions, we carried out a series of 30 minute tests in which we observed the responses of *M. sanctus* to various objects placed in their foraging area. All these objects were unfamiliar to the ants.

At the start of each test we placed 15 objects: 5 halves of rye seeds, 5 small (about 6 mm in diameter) pieces of bread and 5 pieces of rye chaff in the middle part of the foraging area of our colony of *M. sanctus* (see Figure 2). Rye seeds were freshly cut in halves, because in the pilot tests we found out that freshly cut or crushed seeds are much more attractive for *M. sanctus* than entire intact seeds, most probably because they provide more chemical cues. During the subsequent 15 minutes we recorded all acts of pseudostinging directed toward these objects, making a distinction between gaster flexing in the direction of the object and "true pseudostinging" involving touching it with the gaster tip.

In each experimental test, after the initial 15 minutes, we placed a dead housefly (killed by freezing and then allowed to warm) close to other objects already present in the foraging area of the colony. During the next 15 minutes we continued to record all acts of gaster flexing and of true pseudostinging directed to these objects and to the fly. In control tests carried out in the same way as the experimental ones, no fly was added after the initial 15 minutes. If any of the objects offered to the ants at the start of the test was transported by them to the nest chamber, we added a new object of the same category. After the end of each test, we removed all test objects from the foraging area of the colony.

Only one test was carried out on each day. The experimental tests were carried out alternately with the control tests. The tests were carried out between 10:00 and 14:00. We tried to keep the exact hour of the start of each test random. The ambient temperature was fairly constant (21±2°C).

The tests were grouped in 4 series of 5 tests each. Within each series, the tests were carried out on 5 successive days. The successive series of tests were separated by the intervals of 3 days, 17 days and 3 days. Altogether, the first part of Experiment 1 comprised 20 tests (10 experimentals and 10 controls).

The preferences of the ants in orienting their pseudostinging behavior toward various objects present simultaneously in their foraging area were analysed by means of the one-sample χ^2 test. We compared the observed frequencies of a given class of pseudostinging responses) directed towards various objects with the frequencies expected assuming that the ants were distributing these responses uniformly among all objects offered to them during a given test. These expected frequencies of pseudostinging responses were calculated taking into account the fact that during the control tests and during the first half of each experimental test the ants could respond to 15 objects (5 halves of rye seeds, 5 pieces of bread and 5 pieces of rye chaff), whereas during the second half of each experimental test they could respond to 16 objects (15 objects used previously and a housefly).

The changes in the frequency of various classes of pseudostinging behavior as a function of time were analysed by means of a linear regression analysis.

To analyse the effect of the presence of animal prey on the frequency of pseudostinging responses directed by *M. sanctus* toward nonprey items present simultaneously in their foraging area, we compared the data recorded during the second half of each experimental test (in the presence of the housefly) with those recorded during the second half of each control test (in the absence of the housefly). These comparisons were made by means of the Friedman two-way analysis of variance by ranks.

Results and Discussion

1. What objects may trigger pseudostinging behavior in M. sanctus? As can be seen in Figure 3, Test Series 1–4, *M. sanctus* displayed pseudostinging behavior in response to all objects used in our tests: seeds (halves of rye seeds), other food items (pieces of bread), dead insect prey (dead houseflies) and inedible food-associated items (rye chaff).

The majority (51.2%) of pseudostinging responses of *M. sanctus* recorded during the first 4 series of the tests was directed to halves of rye seeds, significantly more frequently than expected (one-sample χ^2 test: $P < 0.0001$). We can thus conclude that *M. sanctus* are more likely to show pseudostinging behavior in response to attractive food than in response to less attractive food items or to inedible objects.

2. Do M. sanctus *show greater readiness to display pseudostinging in response to insect prey than in response to nonprey objects?* Dead

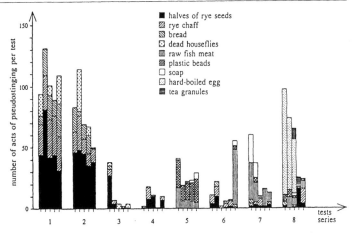

■ halves of rye seeds
▨ rye chaff
▧ bread
⊡ dead houseflies
▥ raw fish meat
▧ plastic beads
□ soap
⊡ hard-boiled egg
▦ tea granules

Figure 3. The results of the Experiment 1. Bars represent the total numbers of acts of pseudostinging (without making the distinction between gaster flexing and true pseudostinging) directed by foragers of Messor sanctus *to various unfamiliar objects placed in their foraging area on successive tests.*

houseflies represented 6.3% of the total number of the objects offered to the ants during the second half of each experimental test. However, the ants directed 101 out of the total number of 550 acts of pseudostinging to the flies during the second half of 10 experimental tests, significantly (one-sample χ^2 test: P < 0.0001) more than expected assuming that they distributed uniformly their pseudostinging responses between all 16 objects present in their foraging area during the second half of each experimental test. We can thus conclude that *M. sanctus* do indeed show greater readiness to display pseudostinging in response to dead houseflies than in response to other objects offered to them in our tests.

3. *Does the presence of insect prey influence the frequency of pseudostinging responses directed by* M. sanctus *to other objects present simultaneously in their foraging area?* To answer that question, we analysed (1) the total number of acts of gaster flexing, (2) the total number of acts of true pseudostinging and (3) the total number of all pseudostinging responses (without making a distinction between gaster flexing and true pseudostinging) directed by *M. sanctus* to all nonprey objects present in their foraging area (rye seeds, rye chaff and pieces of bread) during the second half of each experimental

test, in the presence of the housefly, and during the second half of each control test, in the absence of the housefly.

No significant differences were found in any of these comparisons (Friedman two-way analysis of variance by ranks). We can thus conclude that the presence of insect prey does not exert any modulatory effect (either enhancing or suppressing) on the frequency of pseudostinging responses directed by *M. sanctus* to other objects present simultaneously in its vicinity. Pseudostinging responses of *M. sanctus* are directed in a fairly selective way toward the objects that trigger them.

4. An unexpected finding was the rapid decrease in the frequency of pseudostinging responses as a function of time. As can be seen in Figure 3, the frequency of pseudostinging responses per test decreased quickly as a function of time. Linear regression analyses were carried out in which the dependent variables were: (1) the total number per test of the acts of gaster flexing (GF), (2) the total number per test of the acts of true pseudostinging (PS) and (3) the total number per test of all pseudostinging responses (GF + PS). The independent variable was time. The results of these analyses are shown in Table 1.

As can be seen, with only one exception (gaster flexing responses directed to dead houseflies), the values of the analysed indices showed always a significant decrease as a function of time. Hence, these findings demonstrate that the readiness of *M. sanctus* to show pseudostinging behavior in response to a given class of objects depends not only on the properties of these objects but above all on the degree of the familiarization of the ants with that category of objects. Novel objects trigger more pseudostinging responses than familiar ones. In the course of the familiarization of the ants with a particular class of novel objects, the frequency of both gaster flexing and true pseudostinging decreases rapidly with time.

Experiment 1, Series 5–8: Pseudostinging Behavior as a Response to Novelty

Series 5, Tests 5–8. The decrease in the number of pseudostinging responses per test observed in the first part of the Experiment 1 might have been a consequence of two different processes: (1) a general rise of the threshold for that behavior or (2) a selective habituation of the ants to the objects used in our tests. To discriminate between these two possibilities, we carried out another 4 series of five 30-minute tests, in which we tested the responses of the ants to new classes of unfamiliar objects.

Table 1. The results of the linear regression analysis of the data obtained during the first part of the Experiment 1 (Series 1–4).

Response	Directed to	Independent variable (time)	Slope	Intercept	Correlation Coefficient	Probability
GF	rye seeds	20 tests	–1.9985	39.0842	–0.733857	< 0.001
PS	rye seeds	20 tests	–1.3128	21.3842	–0.803127	< 0.0001
GF + PS	rye seeds	20 tests	–3.3113	60.4684	–0.852868	< 0.00001
GF	rye chaff	20 tests	–0.8902	17.1474	–0.638872	< 0.01
PS	rye chaff	20 tests	–0.9023	13.8737	–0.792110	< 0.0001
GF + PS	rye chaff	20 tests	–1.7925	31.0211	–0.879027	< 0.00001
GF	pieces of bread	20 tests	–0.5474	9.7437	–0.566885	< 0.01
PS	pieces of bread	20 tests	–0.6323	9.8894	–0.732359	< 0.001
GF + PS	pieces of bread	20 tests	–1.1797	19.6368	–0.766370	< 0.0001
GF	dead houseflies	10 tests	–1.3455	14.6000	–0.475114	0.165 (NS)
PS	dead houseflies	10 tests	–1.1091	9.0000	–0.770590	< 0.01
GF + PS	dead houseflies	10 tests	–2.4546	23.6000	–0.661206	< 0.05
GF	rye seeds, rye chaff and pieces of bread	20 tests	–3.4361	65.9789	–0.754400	< 0.001
PS	rye seeds, rye chaff and pieces of bread	20 tests	–2.8474	45.1474	–0.837769	< 0.00001
GF + PS	rye seeds, rye chaff and pieces of bread	20 tests	–6.2835	111.126	–0.904723	< 0.00001
GF	rye seeds, rye chaff, bread and houseflies	20 tests	–3.8241	73.6526	–0.748382	< 0.001
PS	rye seeds, rye chaff, bread and houseflies	20 tests	–3.1444	49.1758	–0.839276	< 0.00001
GF + PS	rye seeds, rye chaff, bread and houseflies	20 tests	–6.9684	123.3680	–0.906308	< 0.00001

Dependent variables: GF = the total number per test of the acts of gaster flexing directed towards a given class of objects; PS = the total number per test of the acts of true pseudostinging directed toward a given class of objects; GF + PS = the total number per test of the acts of both gaster flexing and true pseudostinging directed toward a given class of objects.

Independent variable: time (20 successive tests in all cases except the responses to houseflies, which were recorded only on 10 tests).

The first of these series (Series 5) started 3 days after Series 4 in the preceding part of Experiment 1. It comprised 5 tests. On each test, we offered to the ants 6 new unfamiliar objects: 2 pieces of raw fish meat, 2 plastic beads and 2 pieces of soap. All these objects were round and relatively small (3–6 mm in diameter). The aim of these tests was to determine:

1.   whether pseudostinging behavior of *M. sanctus* may appear in response to inedible objects not associated with food, either relatively odorless (plastic beads) or possessing strong odor (soap);

2.   whether new unfamiliar objects will trigger a significant peak in the frequency of pseudostinging responses per test in relation to the level observed during the last series of the tests with the now familiar rye seeds, rye chaff, pieces of bread and dead houseflies (Series 4).

Results and Discussion. As can be seen in Figure 3, the results of the tests of the Series 5 confirmed that pseudostinging may be triggered both in response to food items (fish meat) and to inedible objects, both relatively odorless (plastic beads) and possessing strong odor not associated with food (soap). Plastic beads triggered even more acts of pseudostinging than pieces of fish meat, possibly because of their seed-like shape (Figure 3).

As can be seen in Figure 3, the ants responded to new unfamiliar objects by increasing the frequency of pseudostinging responses, although now we were offering to the ants only 6 objects instead of the 15 or 16 objects used in the previous tests. The total number of gaster flexing responses per test, the total number of true pseudostinging responses per test and the total number of all pseudostinging responses per test were all significantly higher ($P < 0.05$; Friedman two-way analysis of variance by ranks) during the tests of the Series 5 than during the tests of the preceding Series 4.

These results demonstrate thus that although the frequency of pseudostinging responses directed by *M. sanctus* to unfamiliar objects decreases rapidly with time in the course of the familiarization of the ants with these objects, that behavior reappears in response to new classes of novel objects. These data also show that the decrease in the total number of pseudostinging responses directed to the objects used in the first part of our experiment did not result from the general rise of the threshold for that behavior but from the selective habituation of the ants to these objects.

Series 6, Tests 1–5. To determine whether *M. sanctus* retained the suppression of pseudostinging response to the three categories of objects used during the first part of the Experiment 1 (Series 1–4), in the next series of tests (Series 6), we retested the ants with these objects in a way identical to that during the control tests of Series 1–4, i.e., by placing in their foraging area 5 halves of rye seeds, 5 pieces of rye chaff and 5 pieces of bread but without adding a dead housefly in the second half of the test. These tests started 3 days after the end of the Series 5 and were carried out on five successive days.

Results and Discussion. The values of the following indices for Series 4 and Series 6 were compared (Friedman two-way analysis of variance by ranks):

1.   The total number per test of acts of gaster flexing (GF) directed toward (1) halves of rye seeds, (2) pieces of rye chaff, (3) pieces of bread and (4) all objects offered to the ants (rye seeds, rye chaff and pieces of bread);
2.   The total number per test of acts of true pseudostinging (PS) directed toward the same four categories of objects as stated above; and, lastly,
3.   The total number per test of all pseudostinging responses (GF + PS) directed toward these four categories of objects.

That analysis revealed significant differences between the results of the tests of Series 4 and Series 6 in only 1 out of 16 cases: the frequency of gaster flexing responses directed to pieces of rye chaff proved to be significantly ($P < 0.05$) higher during the tests of Series 6 than during the tests of Series 4. In the remaining 15 cases, no significant differences were found between the results obtained during Series 4 and 6 tests. We may thus conclude that the selective habituation of the pseudostinging response to the objects used in the first part of the Experiment 1 was fairly well retained 10 days after its end, in spite of the fact that in the meantime the ants showed that behavior in response to novel classes of unfamiliar objects.

Series 6, Test 6. Thirty minutes after the end of the fifth test of Series 6 we carried out an additional 30-minute test, in which we observed the responses of the ants to all objects used so far in our tests. We did not remove the objects used in the test that had just terminated, i.e., rye seeds, rye chaff and pieces of bread, and we added the objects belonging to the categories used in Session 5, i. e., pieces of fish meat, plastic beads and pieces of soap. Each of these categories of objects was represented by 2 items.

Results and Discussion. Almost all (94.5%) of the observed acts of pseudostinging were directed to less familiar objects, significantly more than expected (one-sample χ^2 test: P < 0.0001). In other words, pseudostinging behavior of *M. sanctus* was directed preferentially to unfamiliar new objects. The habituation to formerly used "unfamiliar" objects was fully retained, in spite of the fact that they were offered to the ants simultaneously with additional unfamiliar objects. Thus, these data provide further support for our earlier conclusion that pseudostinging responses of *M. sanctus* are directed selectively to the objects which trigger them.

Series 7. To confirm fully these conclusions, 10 days after the end of Series 6 we carried out yet another series of 5 tests (Series 7), in which we continued to observe the responses of the ants to all objects used so far, except the houseflies. This time, each category of objects was represented by 2 items rather than 5. As usual, the tests were carried out on 5 successive days.

Results and Discussion. On the first 2 tests, the ants directed 88.3% and 86.5%, respectively, of their pseudostinging responses to less familiar objects, significantly more than expected (one-sample χ^2 test: P < 0.0001 and P < 0.001, respectively). This confirms our earlier conclusion that pseudostinging behavior triggered by new classes of unfamiliar objects is directed selectively toward these objects. During the subsequent 3 tests, the ants showed few pseudostinging responses to any of the objects offered to them; evidently, by now they had become habituated also to the second group of unfamiliar objects. Nevertheless, on the fourth test they once again responded preferentially to an unfamiliar object, raw fish (one-sample χ^2 test: P < 0.05).

The dynamics of the habituation of *M. sanctus* to the second group of unfamiliar objects (pieces of raw fish meat, plastic beads and pieces of soap) were analyzed quantitatively by means of the linear regression method. In these analyses, the dependent variables were: (1) the total number per test of the acts of gaster flexing (GF), (2) the total number per test of the acts of true pseudostinging (PS) and, lastly, (3) the total number per test of all pseudostinging responses (GF + PS) directed toward a given category/categories of objects. The independent variable was time (16 successive tests).

As revealed by that analysis, the frequency of pseudostinging responses (GF, PS and GF + PS) directed toward the second group of unfamiliar objects used in Experiment 1 showed significant linear decrease with time (Table 2). However, a separate analysis of the dynamics of the habituation of the ants to pieces of raw fish meat, to plastic beads and to pieces of soap revealed that the frequency of pseudostinging responses showed a significant linear decrease with time only in the case of plastic beads.

Series 8, Tests 1–2. To test whether we could generate yet another peak of pseudostinging behavior by offering the ants yet another class of unfamiliar objects, we carried out an additional series of 5 tests on 5 successive days (Series 8). That series of tests started 3 days after the end of Series 7.

On the first 2 tests we again offered the ants 2 items of each category used in the preceding tests with the exception of the housefly and additionally 2 items of a new class of unfamiliar objects: small (about 6 mm) pieces of hard-boiled egg white.

Results and Discussion. Pieces of egg white were both unfamiliar to the ants and highly attractive as food, as evident from their behavior. The ants responded to the egg whites by dissecting them into smaller pieces that were then retrieved to their nest, and they released high peaks of pseudostinging behavior (Figure 3). Pseudostinging behavior was again directed selectively toward the egg white. Pieces of egg received 88.8% and 79.9%, respectively, of the total number of acts of pseudostinging recorded during the first 2 tests of Series 8, significantly more than expected (one-sample χ^2 tests: P < 0.0001 in both analysed cases).

Series 8, Test 3. To test further whether *M. sanctus* show indeed greater readiness to display pseudostinging behavior in response to novel ob-

Table 2. The results of the linear regression analysis of the responses of *M. sanctus* to the second group of unfamiliar objects offered to them during the Experiment 1.

Response	Directed to	Slope	Intercept	Correlation Coefficient	Probability
GF	fish meat	−0.5382	14.325	−0.266102	NS
PS	fish meat	−0.1426	2.775	−0.142647	NS
GF + PS	fish meat	−0.6809	17.100	−0.281056	NS
GF	plastic beads	−0.9647	13.450	−0.716130	< 0.01
PS	plastic beads	−0.9324	3.350	−0.663070	< 0.01
GF + PS	plastic beads	−1.1971	16.800	−0.747582	< 0.001
GF	pieces of soap	−0.0912	2.275	−0.149774	NS
PS	pieces of soap	−0.1412	2.825	−0.195253	NS
GF + PS	pieces of soap	−0.2324	5.100	−0.179808	NS
GF	fish meat, plastic beads and pieces of soap	−1.5941	30.050	−0.635706	< 0.01
PS	fish meat, plastic beads and pieces of soap	−0.5162	8.950	−0.505911	< 0.05
GF + PS	fish meat, plastic beads and pieces of soap	−2.1103	39.000	−0.612734	< 0.05

Dependent variables: GF = the total number per test of the acts of gaster flexing directed toward a given class of objects; PS = the total number per test of the acts of true pseudostinging directed toward a given class of objects; GF + PS = the total number per test of the acts of both gaster flexing and true pseudostinging directed toward a given class of objects.

Independent variable: time (16 successive tests).

jects if they are attractive as food, on the next test we offered the ants all objects used in the previous tests and, in addition, another class of objects: 2 tea granules, objects unfamiliar to ants but inedible.

Results and Discussion. As can be seen in Figure 3, although tea granules elicited several pseudostinging responses (n = 9), they did not trigger a significant peak of pseudostinging behavior comparable to that triggered by pieces of egg white, which supports our conclusion that inedible unfamiliar objects trigger less pseudostinging responses than novel food items.

Series 8, Tests 4–5. Finally, during the last 2 tests we offered the ants all objects used so far, but we removed the pieces of egg white, which were clearly most attractive to them and monopolised the attention of many foragers.

Results and Discussion. Even now tea granules did not trigger any important peak of pseudostinging behavior, although they again triggered several pseudostinging responses (Figure 3). This confirms finally our conclusion that although all kinds of unfamiliar objects may trigger pseudostinging responses in *M. sanctus*, these ants show higher readiness to display pseudostinging behavior in response to attractive food items.

Experiments 2 and 3: Behavioral Processes Underlying the Familiarization of *M. sanctus* with Novel Varieties of Seeds

In Experiment 1 we recorded only pseudostinging responses (acts of gaster flexing and of true pseudostinging). Hence, we could not analyse the possible modifications of other responses of the ants to the objects used in our tests. Moreover, it must be remembered that the responses of the harvester ants to seeds involve whole chains of more elementary behavioral units in which each next step may occur only if the object has not been rejected by the ants at any of the preceding steps (Johnson 1991; Johnson, Rissing & Killeen, 1994; see also Figure 4). Thus, each gaster flexing response must be preceded by a chain of at least three successive behavioral events: approach to the object, antennal contact with it, seizing it in the mandibles. Gaster flexing cannot occur if the object is rejected by the ant at any of these stages. Similarly, each act of true pseudostinging must be preceded by at least four successive events: approach, antennal contact, seizing/biting, gaster flexing.

This implies that the decrease in the frequency of pseudostinging behavior occurring in the course of the familiarization of *M. sanctus* with novel objects observed by us in the Experiment 1 may have resulted from two kinds of processes:

1. specific processes influencing directly the readiness to display gaster flexing and/or true pseudostinging (for instance, by raising the threshold for that behavior, or by its selective inhibition) or

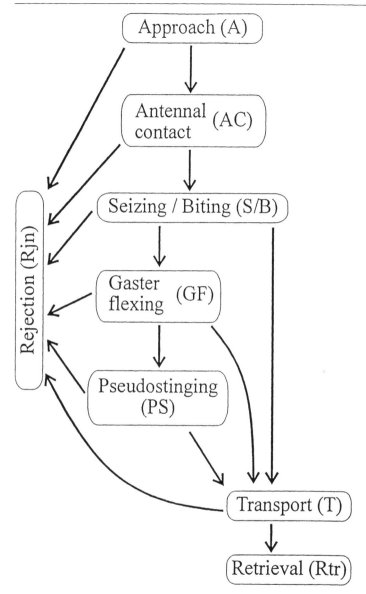

Figure 4. The flow diagram of sequences of behavioral events triggered in workers of Messor sanctus *in response to a seed.*

2. processes exerting a suppressing effect on the frequency of gaster flexing and true pseudostinging via a significant increase in the frequency of rejections of a given class of objects at earlier steps of the sequence of behavioral events that might otherwise lead to gaster flexing/true pseudostinging.

The data obtained in Experiment 1 did not allow us to discriminate between these two possibilities, because in that experiment we recorded only pseudostinging responses. Therefore, in Experiments 2 and 3 we recorded the whole behavioral sequence shown and were able to analyse our data taking into account the frequency of each of the successive steps of these behavioral sequences contingent on the frequency of the response representing the preceding step in that sequence.

In the spring of 1992, our colony of *M. sanctus* was used in two further experiments (Experiments 2 and 3), conducted primarily to find out whether pseudostinging behavior of these ants involves chemical marking of objects touched by them with the gaster tip. These experiments also allowed us to investigate more closely the behavioral processes intervening in the course of familiarization of *M. sanctus* with novel varieties of seeds.

In Experiment 2 we investigated the responses of *M. sanctus* to novel food items: halves of cooked grains of long rice. The experiment was conducted between 24 February and 6 March 1992 and consisted of 2 series of five 1-hour tests. Within each series the tests were carried out on five successive days. The 2 series of tests were separated by a 3-day interval.

In Experiment 3 we investigated the responses of *M. sanctus* to yet another class of unfamiliar food items: halves of sunflower seeds. That experiment was carried out during the period of 20 March to 3 April 1993. Eleven 1-hour tests were carried out. The first 2 tests were separated by a 3-day interval, then a series of 5 tests was conducted on 5 successive days, followed by 5 further tests on 5 successive days after another 3-day interval.

Procedure. In all tests we used freshly cut halves of seeds to make them more attractive for the ants, by providing more chemical cues, to reduce their weight and to facilitate in such a way their handling and transport.

At the start of each test, 2 halves of a seed were placed simultaneously in the middle of the foraging area of our captive colony of *M. sanctus,* at the same place where unfamiliar objects were offered to the ants during Experiment 1 (Figure 2). The seeds were always about 3 cm apart. During the test we recorded all behavioral sequences shown by the ants in response to the seeds: no visible response to the seed; antennal contact(s) with the seed; seizing the seed in the mandibles and/or seed biting; gaster flexing in the direction of the seed; true pseudostinging of the seed; transport of the seed; retrieval of the seed to the nest (see Figure 4). We always noted to which seed the observed behavior was directed.

When both halves had been touched by the ants with their gasters in an act of pseudostinging, we replaced the seed that had been touched first by a fresh one. That procedure was repeated each time the newly added seed had been touched with the gaster by an ant showing pseudostinging behavior. That allowed us to subsequently compare the responses of the ants to seeds that had been and that had not been subject to true pseudostinging. After the end of each test, all the seeds used were removed from the foraging area. Any seed retrieved to the nest was always replaced immediately by a fresh one, so that 2 seeds were always simultaneously present in the middle of the foraging area.

The data obtained in Experiments 2 and 3 were analyzed in two independent ways. In the principal analysis, we investigated the effect of handling and pseudostinging of the seed on subsequent responses to it. The results of those analyses are not reported here for space considerations. Our data suggest that *M. sanctus* are indeed able to discriminate between the seeds that have already been handled by them or by their nestmate and those that have not been handled by any ant. However, the effects of pseudostinging seem to play a much less important role in that phenomenon than the effects of seizing/biting the seed, as shown in as yet unpublished findings of ours.

The results of Experiments 2 and 3 allowed us also to investigate more closely behavioral processes occurring in the course of the familiarization of *M. sanctus* with novel varieties of seeds. To analyze the dynamics of the familiarization of *M. sanctus* with novel varieties of seeds investigated by us in Experiments 2 and 3, we calculated for each test the total number of behavioral sequences that included (1) approach to the seed (ΣA); (2) antennal contact with the seed (ΣAC); (3) seizing and/or biting ($\Sigma S/B$); (4) transport of the seed taking place directly after its being seized or bitten ($\Sigma T_{S/B}$), without being preceded by gaster flexing or by true pseudostinging; (5) gaster flexing (GF) in the direction of the seed (ΣGF); (6) transport of the seed after gaster flexing but without being preceded by true pseudostinging ($\Sigma T_{G/F}$); (7) true pseudostinging of the seed (ΣPS); (8) transport of the seed preceded by true pseudostinging (ΣT_{PS}); (9) transport of the seed ($\Sigma T = \Sigma T_{S/B} + \Sigma T_{G/F} + \Sigma T_{PS}$); and (10) retrieval of the seed to the nest (ΣRtr). The values of these "primary indices" calculated for each test are shown in Tables 3 and 4.

On the basis of these data, we calculated for each test a further set of secondary indices characterizing the behavior of the ants (also shown in Tables 3 and 4). A number of important differences between the responses of *M. sanctus* to rice grains and to sunflower seeds did not allow us to analyze the data obtained in Experiments 3 and 4 in a strictly symmetrical way.

During Experiment 2 (with rice), only 6 cases of transport were observed. In one case, transport was preceded by gaster flexing in the direction of the seed, and in 5 cases it was preceded by gaster flexing and true pseudostinging of the seed. Therefore, we decided to analyze only the frequency of transport contingent on the frequency of gaster flexing ($\Sigma T/\Sigma G/F[\%]$).

During Experiment 2, two behavioral categories were wholly absent: (1) transport of the grain directly after its being seized/bitten (TS/B) and (2) retrieval of the grain to the nest (Rtr). Therefore, for Experiment 2 we calculated only the following secondary indices: (1) $\Sigma AC/\Sigma A[\%]$; (2) $\Sigma S/B/\Sigma AC[\%]$; (3) $\Sigma G/F \ \Sigma S/B[\%]$; (4) $\Sigma PS/\Sigma G/F[\%]$; (5) $\Sigma G/F \ \Sigma A[\%]$; (6) $\Sigma PS/\Sigma A[\%]$; (7) $\Sigma T/\Sigma G/F[\%]$ (see Table 3).

For Experiment 3, we additionally calulcated (1) $\Sigma T_{S/B}/\Sigma S/B[\%]$; (2) $\Sigma T_{G/F}/\Sigma GF[\%]$; (7) $\Sigma T/\Sigma S/B[\%]$; (8) $\Sigma Rtr/\Sigma T[\%]$. We did not analyze the modifications of the values of the index $\Sigma T_{PS}/ \Sigma PS[\%]$, as only 4 cases of transport of a seed preceded by its true pseudostinging were observed during all of Experiment 3 (see Table 4).

Table 3. The responses of *Messor sanctus* to grains of rice recorded during ten 1-hour tests (Experiment 2).

Indices	Tests										
	1	2	3	4	5	6	7	8	9	10	
				a. primary indices							
ΣA	19	19	31	40	40	23	58	27	38	51	
ΣAC	12	13	16	16	27	14	17	4	24	18	
ΣS/B	7	8	6	4	7	3	2	1	1	3	
ΣGF	6	4	3	4	4	1	2	1	0	3	
ΣT$_{GF}$	0	0	0	1	0	0	0	0	0	0	
ΣPS	4	3	3	2	4	1	0	1	0	1	
ΣT$_{PS}$	2	1	0	2	0	0	0	0	0	0	
ΣT(ΣT$_{GF}$ + ΣT$_{PS}$)	2	1	0	3	0	0	0	0	0	0	
				b. secondary indices							
ΣAC/ΣA [%]	63.2	68.4	51.6	40.0	67.5	60.9	29.3	14.8	63.2	35.3	NS
ΣS/B/ΣAC [%]	58.3	61.5	37.5	25.0	25.9	21.4	11.8	25.0	4.2	16.7 ↘	***
ΣGF/ΣS/B [%]	85.7	50.0	50.0	100.0	57.1	33.3	100.0	100.0	0	100.0	NS
ΣPS/ΣGF [%]	66.7	75.0	100.0	50.0	100.0	100.0	0	100.0	0	33.3	NS
ΣGF/ΣA [%]	31.6	21.1	9.7	10.0	10.0	4.3	3.4	3.7	0	5.9 ↘	***
ΣPS/ΣA [%]	21.1	15.8	9.7	5.0	10.0	4.3	0	3.7	0	5.6 ↘	****
ΣT/ΣGF [%]	33.3	25.0	0	75.0	0	0	0	0	0	0	NS

For the definitions of the indices, see the text and Figure 4.

    No case of transport of a grain directly after its seizing/biting (Ts/B) and of retrieval of a grain to the nest (Rtr) was recorded during this experiment.

    Statistics: Results of the linear regression analysis in which the values of the secondary indices (dependent variables) were regressed against the successive tests (independent variable). Probability level: NS: P > 0.05; ***: P < 0.01; ****: P < 0.001.

The data obtained in Experiments 2 and 3 were analyzed by means of a linear regression analysis, in which the values of the secondary indices were regressed against time (10 successive tests in the case of the Experiment 1, 11 successive tests in the case of the Experiment 3) (Tables 3 and 4).

Results and Discussion of Experiments 2 and 3
Experiment 2: Familiarization of M. sanctus with Grains of Rice.
The results of Experiment 2 are shown in Table 3. As can be seen, in

Table 4. The responses of *Messor sanctus* to sunflower seeds recorded during eleven 1-hour tests (Experiment 3).

Indices	Tests											
	1	2	3	4	5	6	7	8	9	10	11	
a. primary indices												
ΣA	143	42	139	54	55	87	186	128	95	130	170	
ΣAC	114	33	84	28	43	63	163	104	67	116	136	
$\Sigma S/B$	53	7	15	14	23	32	60	48	30	58	47	
$\Sigma_{S/B}$	6	0	1	0	4	1	13	17	13	26	10	
ΣGF	36	7	11	12	17	27	38	19	7	25	21	
ΣT_{GF}	7	0	3	1	4	4	23	12	4	18	8	
ΣPS	7	3	5	7	3	8	1	2	1	1	1	
Σ_{PS}	2	1	0	0	0	1	0	0	0	0	0	
$\Sigma T(\Sigma T_{SB} + T_{GF} + T_{PS})$	15	1	4	1	8	6	36	29	17	44	18	
ΣRtr	6	1	2	1	0	3	18	12	12	23	11	
b. secondary indices												
$\Sigma AC/\Sigma A$ [%]	79.7	78.6	60.4	51.9	78.2	72.4	87.6	81.3	70.5	81.9	80.0	NS
$\Sigma S/B/\Sigma AC$ [%]	46.5	21.2	17.9	50.0	53.5	50.8	36.8	46.2	44.8	50.0	34.6	NS
$\Sigma GF/\Sigma S/B$ [%]	54.7	57.1	46.2	35.7	60.9	59.3	61.7	35.4	20.0	41.4	44.7	NS
$\Sigma T_{S/B}/\Sigma S/B$ [%]	11.3	0	6.7	0	17.4	3.1	21.7	35.4	43.3	44.8	21.3	↗ ***
$\Sigma PS/\Sigma GF$ [%]	19.4	42.9	45.5	58.3	17.6	29.6	2.6	10.5	14.2	4.0	4.8	↘ **
$\Sigma T_{G/F}/\Sigma GF$ [%]	19.4	0	27.3	8.3	23.5	14.8	60.5	63.2	57.1	72.0	38.1	↗ ***
$\Sigma T/\Sigma S/B$ [%]	28.3	14.3	26.7	7.1	34.8	18.8	60.0	60.4	56.7	75.9	38.3	↗ **
$\Sigma Rtr/\Sigma T$ [%]	40.0	100.0	50.0	100.0	0	50.0	50.0	63.2	70.6	52.3	61.1	NS

For the definitions of the indices, see the text and Figure 4.

    Statistics: Results of the linear regression analysis in which the values of the secondary indices (dependent variables) were regressed against the successive tests (independent variable). Probability level: NS: $P > 0.05$; **: $P < 0.025$; ***: $P < 0.01$. For further details, see the text.

the course of their familiarization with grains of rice, *M. sanctus* significantly modified their responses to that variety of potential food items. The most important behavioral modification of the ants consisted in a significant decrease of the readiness to seize/bite the seeds.

As in Experiment 1, we observed a significant decrease in the frequency of gaster flexing and of true pseudostinging as a function of time. However, whereas the frequency of gaster flexing and of true pseudostinging contingent on the frequency of approaches showed a significant decrease with time, the frequency of gaster flexing contingent on the frequency of seizing/biting, and the frequency of true pseudostinging contingent on the frequency of gaster flexing did not decrease significantly with time (Table 3). This suggests that the observed decrease in the frequency of gaster flexing and of true pseudostinging contingent on the frequency of approaches was not caused by specific processes directly influencing the readiness to display pseudostinging behavior but appeared as a secondary effect of the significant increase in the frequency of early rejections of grains of rice (after an antennal contact with the grain).

As was already mentioned, we observed only 6 cases of transport of a grain during Experiment 2. That behavior was observed only on the first 4 tests and then it disappeared altogether (Table 3). It never led to the retrieval of the rice grain to the nest.

The results of Experiment 2 show that in the course of the familiarization with a new variety of seeds, *M. sanctus* may learn to reject these seeds. The principal behavioral modifications observed in that experiment, the decrease in the readiness to seize/bite the grain of rice and the disappearance of the tendencies to transport, were both in the direction of increased readiness to reject grains of rice.

Experiment 3: Familiarization of M. sanctus *with Sunflower Seeds.* The results of Experiment 3 are shown in Table 4. As in Experiment 2, the responses of *M. sanctus* to a novel variety of seeds showed several significant modifications as a function of time. However, this time the ants modified their behavior in the direction of increased readiness to accept novel seeds. The main modifications of behavior of the ants consisted in a significant increase of the readiness to transport the seed directly after its seizing/biting, and in a significant readiness to transport the seed after an act of gaster flexing in its direction. At the same time, we observed a significant decrease in the frequency of true pseudostinging contingent on the frequency of gaster flexing. Thus, this time the decrease in the frequency of true pseudostinging seems to have been caused by some specific processes influencing directly the readiness to display that behavior.

The results of Experiment 3 also suggest that the behavior of the ants does not change only as a function of the number of tests during which they could familiarize themselves with the sunflower seeds. The temporal schedule of the tests with sunflower seeds was somewhat irregular: as a rule, the tests were separated by a one-day interval, but the intertest interval after the first and then after the sixth test was longer (3 days). As clearly seen in Table 4, the data obtained during the last 5 tests (after a 3-day pause in the tests) are strikingly different from those obtained during the first 6 tests. In particular, there is a dramatic increase in the frequency of transport, both occurring directly after seizing/biting the seed and preceded by an act of gaster flexing in its direction. These data suggest that a longer intertrial interval might have facilitated the ants to switch from excited behavior triggered in response to novelty to rapid, unhesitating responses to seeds leading directly to their harvesting.

The Main Conclusions of Experiments 2 and 3 may be summarized as follows:

1.   Harvester ants of the species *M. sanctus* rapidly modify their responses to novel seeds in the course of their familiarization with them.
2.   These modifications may lead in two opposite directions: increased readiness to reject a given variety of seeds or increased readiness to accept that variety of seeds.

In the first case (Experiment 2) the ants learn rapidly to reject the new variety of seeds already after an antennal contact, without seizing them. The tendencies to transport disappear altogether. The frequency of pseudostinging behavior (gaster flexing and true pseudostinging) decreases significantly, but that effect seems to be only a direct consequence of the decrease of the readiness to seize/bite the seeds. It does not seem to be related to any specific modifications of the threshold for the gaster flexing/true pseudostinging behavior.

In the second case (Experiment 3) the modifications of the responses of the ants to novel seeds involve mainly an increase in the readiness to transport the seed. At the same time, the readiness for the true pseudostinging behavior decreases, this time, apparently, in a specific way. The ants shift from behavior dominated by general excitement (manifesting itself in frequent pseudostinging) to direct tendencies to transport following immediately the detection of the seed. In other words, the ants are switching from excited behavior triggered in response to novelty to unhesitating approach-acceptance responses allowing efficient foraging under stable familiar environmental conditions.

General Discussion
The Classical Observations and Hypotheses of Goetsch: Is Pseudostinging a Technique Facilitating Seizing of Seeds?

Responses to seeds strikingly similar to pseudostinging behavior observed by us in *M. sanctus* were described earlier by Goetsch (1928, 1929) in his classical papers on seed-collecting behavior of the ants of the genus *Messor*. According to Goetsch, the attempts of these ants to properly seize a seed often involve an act of forward gaster flexing employed by them to push the seed into place with the gaster tip (Figure 5). Goetsch also proposed that during that act, the ant may simultaneously mark the seed with some chemical secretion providing chemical cues facilitating the subsequent acceptance of the seed by nestmates. As demonstrated by Goetsch (1928, 1929) and subsequently confirmed by Baroni-Urbani (1987), ants of the genus *Messor* often employ a relay during retrieval of food to the nest. Seeds are frequently dropped by their finders near the the trail only to be carried off by other ants. Chemical marking of the seeds might facilitate their redetection by nestmates picking them up along the trail.

Our research confirms the observations of Goetsch; however, we cannot fully agree with his interpretation of the observed phenomena. Even if gaster flexing may sometimes help the ant to properly seize a small transportable object, this does not seem to us to be the primary function of that behavior. Pseudostinging movements of the gaster in the direction of the object seized in the mandibles often do not lead at all to touching of that object with the gaster tip. They are often followed by immediate abandoning of the object, without any tendency to transport. Moreover, we observed that vio-

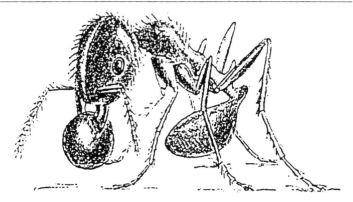

Figure 5. A drawing of Goetsch (1928) showing a large worker of Messor *in train of flexing the gaster in the direction of a seed held in its mandibles. Published originally in* Zeitschrift für Morphologie und Ökologie der Tiere, 10: *353–419. Reprinted by permission of Springer-Verlag New York, Inc.*

lent and prolonged pseudostinging responses may be directed not only to small transportable objects but also to large, untransportable objects such as large food items, forceps, scissors or even the edge of the bowl or the Petri dish in which the ant is confined.

The results of our Experiment 2 (in preparation) strongly suggest that handling of the seed by an *M. sanctus* worker may have a long-lasting effect facilitating its redetection by nestmates. However, seizing the seed with the mandibles seems to play the primary role in that phenomenon; even if the ants deposit some secretion on the seed during pseudostinging, that marking seems to play a secondary role in facilitating redetection of that object by other ants.

Does Pseudostinging Involve the Release of Volatile Alarm Pheromones?

The presence of volatile alarm substances in the abdominal secretions of the ants of the genus *Messor* was reported by Goetsch (1934). Further research of Maschwitz (1964) and Hahn and Maschwitz (1980, 1985) demonstrated that the poison gland secretion of these ants contains volatile alarm pheromones. During our observations of pseudostinging behavior of *M. sanctus*, we did not observe the release of any drops of liquid. However, myrmicine ants may release the contents of their poison gland directly in the air (Hölldobler, Stanton & Markl, 1978); thus, we cannot exclude the possibility that pseudostinging behavior involves the release of volatile pheromones of the poison gland in the air. Whether the acts of pseudostinging are followed by significant modifications of the behavior of the nestmates in the vicinity of the ants carrying out that behavior, in particular, whether they are followed by the short-range alarm recruitment remains to be investigated.

Main Conclusions

The function of pseudostinging behavior is still not fully clear to us. In particular, it remains still to be determined whether pseudostinging involves the release of some chemical secretion(s) or is simply the so called "vestigial behavioral pattern" homologous to true stinging behavior but not fulfilling anymore any adaptive function. However, we are already able to throw some light on the proximate causal factors for that behavior. In particular, we found out that the pseudostinging behavior of the ants of the species *Messor sanctus* Forel may be triggered in response to a wide range of objects, edible as well as inedible ones. We also demonstrated that pseudostinging is directed selectively to the object that had triggered it: objects triggering numerous pseudostinging responses do not enhance the frequency of acts of pseudostinging directed to other objects present simultaneously in their vicinity.

We were also able to show that the readiness of *M. sanctus* to show pseudostinging behavior in response to a given class of objects depends not only on the properties of these objects (attractive food items triggering more pseudostinging responses than clearly inedible objects) but, above all, on the degree of the familiarization of the ants with a given class of objects. As a rule, in the course of the familiarization of the ants with a particular category of novel objects, the frequency of acts of true pseudostinging and of gaster flexing directed to these objects decreases rapidly with time. However, new classes of unfamiliar objects trigger again pseudostinging and gaster flexing responses directed selectively toward these objects.

We also observed that workers of *M. sanctus* display pseudostinging behavior in response to novel varieties of seeds. In two series of laboratory tests, we analyzed more closely behavioral modifications occuring in the course of the familiarization of *M. sanctus* with novel varieties of seeds. We found out that these modifications may lead in two opposite directions: the ants may either learn to reject a given variety of seeds or learn to accept these seeds readily and unhesitatingly.

In the first case (learning to reject a given variety of seeds) the main modification in the behavior of the ants consisted in a significant decrease of the readiness to handle the seeds (seize/bite them). The frequency of gaster flexing and true pseudostinging decreased also significantly as a function of time. However, that decrease was not selective: it reflected simply a significant decrease in the tendency to handle these seeds. The tentatives of transport of the seeds were very unfrequent even at the start of the experiment, and very rapidly they disappeared altogether.

In the second case (learning to accept a given variety of seeds) the main modifications of the responses of the ants to novel seeds consisted in a significant increase of the readiness to transport the seed directly after its seizing/biting. The readiness to transport the seed after an act of gaster flexing in its direction increased, too. At the same time, we observed a significant decrease of the readiness of the ants to display true pseudostinging responses after seizing/biting of the seed. Interestingly, the readiness to carry out an act of gaster flexing did not decrease with time.

Our present data demonstrate thus that in the course of learning to accept a given variety of seeds, *M. sanctus* shift progressively from behavior dominated by general excitement (manifesting itself in frequent pseudostinging) to direct tendencies to transport following immediately the detection of the seed. In other words, the ants are switching from excited behavior triggered in response to novelty to unhesitating approach-acceptance responses allowing efficient foraging under stable, familiar environmental conditions.

The Role of Acquired Factors in the Foraging Behavior of the Harvester Ants

The role of acquired factors in the foraging behaviour of the harvester ants was demonstrated by Rissing (1981) in workers of *Pogonomyrmex rugosus*. As shown by Rissing, these ants learn quickly to specialize on a single species of grass seeds. These individual specializations hold for at least several days, but can also be shifted rapidly. Similar phenomena occur also in another species of harvester ants, *Messor (= Veromessor) pergandei* (Johnson, 1991; Johnson et al.,

1994). The studies of Johnson (1991) and Johnson et al. (1994) demonstrate that learning and memory are important components in diet and patch selection in harvester ants. Our present results confirm the conclusions of Johnson (1991) and Johnson et al. (1994) concerning the importance of learning and memory for the foraging behavior of harvester ants. Our data also confirm that the modifications of the behavior of the ants in the course of their familiarization with novel seeds occur at many independent levels. Additionally, we demonstrated that the ants may learn not only to accept a new variety of seeds but also to reject it.

Johnson (1991) also found that in the course of the familiarization with a new seed patch, the foraging efficiency of workers of *M. pergandei* and *P. rugosus* increases as a result of the modifications of the search behavior of outgoing foragers. The ants learn to visit a more distant source of seeds of known location rather than to search for nearer sources of seeds. As a result, time needed to carry out a foraging trip decreases significantly. Our present data demonstrated that in the course of the familiarization with novel seeds, the foraging efficiency of *M. sanctus* increases as a result of similar switching from hesitating behavior, that is, excited behavior that appears in these ants in response to novelty, to unhesitating, smooth sequences of approach-seizing-transport.

The Relevance of Our Research for Theoretical Ideas of T. C. Schneirla

Our present data support and extend some of T. C. Schneirla's conclusions concerning the mechanisms of complex learning processes in the ants. At the same time, they have also some relevance for his approach-withdrawal theory of behavior.

As shown by Schneirla (1929, 1941, 1943), learning of complex mazes by the ants involves several distinct successive stages. The first stage (called by Schneirla the "initial period of generalized learning") involves the general habituation of the ant to a novel experimental situation. The process of very gradually mastering local choice-points starts only when the ants reach the next stage of complex maze learning, no longer dominated by novelty responses. The existence of these two stages in maze learning in the formicine ants has been subsequently confirmed by Vowles (1964, 1965).

Our data demonstrate that similar phenomena are observed during the familiarization of harvester ants of the genus *Messor* with novel varieties of seeds. During their first tests with previously unknown seeds, these ants show mainly excited behavior, characterized by high frequency of pseudostinging. Only after a few tests do they start to respond to novel seeds by rapid, unhesitating approach and acceptance.

Our findings demonstrate that the initial stage of general habituation to novelty plays an important role not only in the acquisition of maze habits in the formicine ants but also in the acquisition of food preferences in the harvester ants of the genus *Messor*. Our data thus support the conclusion of Schneirla concerning the existence of an "initial period of generalized learning" in complex learning processes in ants and extend its generality to another type of ant learning. Our data also suggest that the necessity of passing through the stage of the "initial period of generalized learning" in the course of the familiarization with novel food items may constitute more than a negligeable constraint on the efficiency of ant foraging.

Our results have also some relevance for Schneirla's approach-withdrawal (A-W) concept of behavior (Schneirla 1939, 1959, 1965). In that concept, Schneirla drew the distinction between proposed A-processes, defined as low-threshold mechanisms that underlie and facilitate actions of approach or seeking, and W-processes, defined as processes related to disruptive or tensional conditions basic to withdrawal or avoidance reactions. Schneirla postulated that low or decreasing stimulative effects can selectively arouse the processes of the A-system, and, conversely, high or increasing stimulative effects can selectively arouse the processes of the W-system (Schneirla, 1965).

Our research on the behavioral processes occurring during the familiarization of *M. sanctus* with novel attractive seeds demonstrated that, in the course of that process, the ants shift progressively from excited, hesitating responses of the W-type, most frequently terminated by the final withdrawal of the ant from the unfamiliar object, to calm, unhesitating responses of the A-type, involving approach to the object and terminated by its acceptance (transport and retrieval to the nest).

Characteristically, A-type behavior appears when the ants start to be habituated to formerly unfamiliar food items. Does it appear in response to "decreasing stimulative effects" as predicted by

Schneirla? Our present data do not allow us to answer that question; it remains open for future investigations.

Acknowledgments

We are very indebted to Alain Lenoir for his participation in the initial part of the research, constant encouragement and many stimulating discussions. We are extremely indebted to Ethel Tobach for her interest in our work, for pointing out its relevance for the theoretical ideas of T. C. Schneirla and for her most stimulating critical comments on the first version of the manuscript. We are very thankful to Gary Greenberg for reading the manuscript and helping to improve the English. We are very grateful to Ullrich Maschwitz and J. O. Schmidt for their most helpful discussions of the role of venom in the behavior of the ants of the genus *Messor*. We thank Alicja Kapuścińka and Maria Kieruzel for technical assistance in rearing the ants and for their help in running a part of the tests.

References

Baroni-Urbani, C. (1987). Comparative feeding strategies in two harvester ants. *Proceedings of the 10th International Congress IUSSI, Munich*, 509–510.

Cerda, X., & Retana, J. (1994). Food exploitation patterns of two sympatric seed-harvester ants *Messor bouvieri* (Bond.) and *Messor capitatus* (Latr.) (Hym., Formicidae) from Spain. *Journal of Applied Entomology, 117*, 268–277.

Gillon, D., Adam, F., & Hubert, B. (1984). Production et consommation de graines en milieu sahélo-soudanien au Sénégal: les fourmis, *Messor galla. Insectes Sociaux, 31*, 51–73.

Goetsch, W. (1928). Beiträge zur Biologie körnersammelnder Ameisen. *Zeitschrift für Morphologie und Ökologie der Tiere, 10*, 353–419.

Goetsch, W. (1929). Untersuchungen an getreidesammelnden Ameisen (Körnerverwertung, Benachrichtigung und Arbeitsteilung im *Messor*-Nest). *Naturwissenschaften, 17*, 221–226.

Goetsch, W. (1934). Untersuchungen uber die Zusammenarbeit im Ameisenstaat. *Zeitschrift für Morphologie und Ökologie der Tiere, 28*, 319–401.

Hahn, M., & Maschwitz, U. (1980). Food recruitment in *Messor rufitarsis. Naturwissenschaften, 67*, 511–512.

Hahn, M., & Maschwitz, U. (1985). Foraging strategies and recruitment behaviour in the European harvester ant *Messor rufitarsis* (F.). *Oecologia, 68*, 45–51.

Hölldobler, B., & Wilson, E. O. (1990). *The ants.* Berlin: Springer-Verlag.

Hölldobler, B., Stanton, R. C., & Markl, H. (1978). Recruitment and food-retrieving behaviour in *Novomessor* (Formicidae, Hymenoptera). I. Chemical signals. *Behavioral Ecology and Sociobiology, 4*, 163–181.

Johnson, R. A. (1991). Learning, memory and foraging efficiency in two species of desert seed-harvester ants. *Ecology, 72*, 1408–1419.

Johnson, R. A., Rissing, S. W., & Killeen, P. R. (1994). Differential learning and memory by co-occurring ant species. *Insectes Sociaux, 41*, 165–177.

López, F., Serrano, J. M., & Acosta, F. J. (1992). Intense reactions of recruitment facing unusual stimuli in *Messor barbarus* (L.). *Duetsche Entomologische Zeitschrift, 39*, 135–142.

Maschwitz, U. (1964). Gefahrenalarmstoffe und Gefahrenalarmierung bei sozialen Hymenopteren. *Zeitschrift für Vergleichende Physiologie, 47*, 596–655.

Maschwitz, U. (1975). Old and new chemical weapons in the ants. In C. Noirot, P. E. Howse & G. Le Masne (eds.), *Pheromones and defensive secretions in social insects* (pp. 41–45). Dijon: Université de Dijon.

Rissing, S. W. (1981). Foraging specializations of individual seed-harvester ants. *Behavioral Ecology and Sociobiology, 9*, 149–152.

Schneirla, T. C. (1929). Learning and orientation in ants. *Comparative Psychology*

Monographs, 6, (Whole No. 4).

Schneirla, T. C. (1939). A theoretical consideration of the basis for approach-withdrawal adjustments in behaviour. *Psychological Bulletin, 37,* 501–502.

Schneirla, T. C. (1941). Studies on the nature of ant learning: I. The characteristics of a distinctive initial period of generalized learning. *Journal of Comparative Psychology, 32,* 41–82.

Schneirla, T. C. (1943). The nature of ant learning: II. The intermediate stage of segmental maze adjustment. *Journal of Comparative Psychology, 35,* 149–176.

Schneirla, T. C. (1959). An evolutionary and developmental theory of biphasic processes underlying approach and withdrawal. In M. R. Jones (ed.), *Nebraska symposium on motivation. Vol. 7* (pp. 1–42). Lincoln: Univ. Nebraska Press.

Schneirla, T. C. (1965). Aspects of stimulation and organization in approach-withdrawal proceses underlying vertebrate behavioral development. In D. S. Lehrman, R. Hinde & E. Shaw (eds.), *Advances in the study of behavior, Vol. 1* (pp. 1–74). New York: Academic Press.

Sheata, M. N., & Kaschef, A. H. (1971). Foraging activities of *Messor aegyptiacus* Emery. *Insectes Sociaux, 18,* 215–226.

Vowles, D. M. (1964). Olfactory learning and brain lesions in the wood ant *(Formica rufa). Journal of Comparative Psychology, 58,*105–111.

Vowles, D. M. (1965). Maze learning and visual discrimination learning in the wood ant *(Formica rufa). British Journal of Psychology, 56,* 15–31.

Contributors

Charles I. Abramson
Department of Psychology
Oklahoma State University

Thomas M. Alloway
Erindale Campus
University of Toronto

G. M. Dlussky
Moscow State University
Moscow, Russia

Niles Eldredge
Department of Invertebrates
The American Museum of
  Natural History

Ewa Joanna Godzińska
Laboratory of Ethology
Nencki Institute of Experi-
  mental Biology
Warsaw, Poland

Deborah M. Gordon
Department of Biological
  Sciences
Stanford University

Jerry Hirsch
Departments of Psychology
  and of Ecology, Ethology,
  and Evolution
University of Illinois at
  Urbana-Champaign

Rudolf Jander
Division of Biological Sciences
University of Kansas

Julita Korczyńska
Laboratory of Ethology
Nencki Institute of
  Experimental Biology
Warsaw, Poland

Robin J. Stuart
Department of Entomology
Rutgers University

Anna Szczuka
Laboratory of Ethology
Nencki Institute of
  Experimental Biology
Warsaw, Poland

Michael G. Terry
Committee on Biopsychology
University of Chicago

Howard Topoff
Department of Psychology
Hunter College of CUNY, and
Department of Entomology
The American Museum of
    Natural History

James F. A. Traniello
Department of Biology
Boston University

Adrian Wenner
Department of Biological
    Sciences
University of California at
    Santa Barbara

Species Index

Italic page numbers indicate illustrations.

Name Index

Italic page numbers indicate illustrations.

Subject Index

Italic page numbers indicate illustrations.

Printed and bound by CPI Group (UK) Ltd, Croydon, CR0 4YY

22/10/2024

01777615-0007